Eli Saida

God and the Illusion of Time©

Published by Eli Saida, 2013.

Printed by CreateSpace, an Amazon Company

United States Copyright Office
Library of Congress

Saida, Eli, 2014—

God and the Illusion of Time

ISBN 9 781492 388494 (trade pbk.)

RN: TXu001631571/ 4/15/2008

RN: TXu1-916-763 3/25/2014

You can purchase this book in paperback or E-book from Amazon.

The birthplace of this book was in the United States of America, Planet Earth, Solar System #1, Milky Way Galaxy, God's Little Universe.

I dedicate this book to my son Eric; for his life accomplishments and for being a good soul and a good father.

Thanks to all my editors, and thank you God. I couldn't have accomplished this book without you.

"God and the Illusion of Time"
In memory of the ones who left us and went to Heaven.

Introduction

The Newest God Bible.

God's own version about the creation of our universe and our matter.

God explains: the laws of nature; what is gravity; what is time & why is it an illusion; the principle of time travel; why is the speed of light always constant and relative to what; what are black holes and how to build one; what is evolution; the creation of the first gene; where did we come from and why do we have the need to believe in a higher power; where did God come from and what does he have to say about "intelligent design." What is knowledge? What is the meaning of life? A deep look at the infinite and timeless universe which is around our limited and temporary universe, to the smallest unit of matter - the "photon".

This book is amazing. It contains information that has never been seen on Earth before. This book was written with no help or influence from anybody or any book (except for Einstein) only with God's help, and it took ten years to create. Everything in this book is original. The information in this book will satisfy, help and enlighten many scientists. This book will bolster Atheist beliefs that there is no real physical God and on the other hand will help the average believers understand their imaginary God much better, since illusion and imagination can seem very real and powerful. In fact, imagination is the most powerful force in our universe.

This book is not for everyone. You have to be at least 21 years young, mature, with some knowledge and passion toward science. If you are old enough to read and understand the old Bible, you certainly can read this Bible since it was written with the help of an older and more mature God.

This book reveals many secrets about our universe and the power of the mind, It shows different ways of thinking and contains some black humor.

The author had to repeat himself many times in different chapters because the information is new and it's the first time seen on Earth.

As a true God's Bible this book has all the answers in principle. If you still have any questions after you read this book we will gladly answer your questions. Follow us on Twitter@ eli_saida and Facebook@ God and the Illusion of Time

About the Author

The author was born on Earth (an Earthling) in the year 1950, son of a father from Mars and a mother from Venus. The mother gave birth to nine children, which the author was placed in the middle. An Illiterate mother that spent her entire life working like a slave in the kitchen: from early morning to late at night she prepared food from scratch and cared for her large family with love and devotion.

A very intelligent father that always sought knowledge without the ability to expand his education as he had to work long and hard hours to provide for his large family. He was controlled by his subconscious, and would lose his good temper from time to time. He worshiped God as he understood him by the rules of his religion.

The author attended God of knowledge college since 1950 and received SPHD (The S stands for Super PHD) in God knowledge from God. Hand-picked by God (God of knowledge wanted human robot # 18,000,000,000 to write this book but he was too blinded by religion and I human robot # 18,000,000,001 became the chosen one) to deliver his message of understanding and knowledge about the basics of our universe.

Not coming from the academic forest, the author's vision wasn't blinded by the trees and brain washed by old beliefs. The author was able to see further and deeper into our universe by seeing what makes us tick-by out of the box thinking. As he was reading natures hard to follow patterns, only with God's help as God revealed all the secrets of our universe in principles while the author was writing this book.

Table of Contents

1. What's Going On?

Throughout the history of mankind almost all human beings have believed in some kind of spirit, god or supernatural force that controls life, death and everything else in the universe.

These days most people believe in some kind of divine force. They participate in many different religions, each person believing that their religion is the right one. Some believe that their faith is the only one and that the others are wrong and sometimes literally dead wrong. These beliefs come from deep within the human being and are different in their intensity from human to human and from time to time. Some human beings are ready to die for what they believe and some will have the need to pray many times a day or to give as much as they can to religious causes.

If you separate identical twins and teach them two completely different religions at an early age, each of them will believe the religion they were taught is the right one and that the other religion is wrong.

So, what's going on? Can two different religions be right?

We know that some people follow the rules and laws of their religion. They help other people (especially their own) and follow their leader. However, all these people, including the leaders, are subject to the same future as people from other religions or people that do not follow any religion.

All people have the same problems regardless of their religious beliefs or lack of beliefs. All people get sick or have an accident from time to time. They all have to provide food to eat, they all need air to breath and they all will die. In short, they all follow the laws of nature.

How come religious people who believe in God do things that are contrary to their religious beliefs? Some even lie, steal, rape and murder.

What's going on?

Some people might tell you that if you do not believe in God or do not follow their way of thinking something bad will happen to you. The truth of the matter is that it is almost certain something bad will happen to you. You might get sick, have an accident or encounter natural forces like earthquakes and extreme weather. These are part of the cycles of nature. They destroy, they bring new life and they are inevitable. And it is an unquestioned certainty that you will die sometime in the future.

Religious leaders try to control people with the fears that are genetically embedded in all of us. They promise a better life to those who follow their specific rules and religious beliefs. In reality, a better life comes to a lot of people from different backgrounds through hard work and serendipity (being in the right place at the right time). The leaders even promise that following their religious rules and commands will offer a better life after death, and say that not following them will result in going to Hell. How do they know that? Have they come back from the dead? Do they have supernatural powers? Do they not have the need to eat food or breathe air like us? Where do they get the knowledge to say all the things they say? Where do we get the beliefs that we're never going to die or that our souls will live forever?

When something bad happens to people who believe in God or to their loved ones, such as the loss of home, money or a job, sickness, an accident or even sudden death, they can become confused. Sometimes they stop believing, try to find answers in a different religion or say that God is great and his ways are mysterious.

People believe that God is like a superhuman being, that He (who could be a She or a sexless entity) is immortal, strong and all-knowing. Yet, like a human leader, He also gets mad if you do not follow Him and do what He wants you to do, if you do not pray for Him and say how great He is. You must bow to Him because He is the greatest King of Kings. You have to follow his rules. It is only in this way that you might live peacefully in His mercy and knowledge and have a decent life on Earth. If you are a good person and follow His rules, you might have a wonderful life even after you die.

It is true that human beings do feel better if they believe in God or something supernatural on which they can depend in bad times, and we all have bad times sometimes. It is fundamental, it is in our genes that we need somebody powerful, a leader to direct us in life, to find out right from wrong, because life is very complicated and many things can and do go wrong. Life can take many twists and turns with uncertainty built in as a feature of everyday living. You will never have all the necessary information all the time to predict events in the future. You have only probabilities. So it is comforting to depend on something that has the knowledge and the power to see the future, to control life and death, and for that matter, the whole universe.

God does not obey the laws of nature, so we feel comfort believing in Him. We feel better mentally, we have less anxiety about life, and we know this supernatural being can help us in bad times and in making decisions in life. We hope He can give us immortal life and independence from the laws of nature. For these reasons we feel better belonging to something powerful.

This belief in a higher power, which comes from deep within the human being, relaxes the mind and makes the physical body function better. If older people believe in God and attend religious groups, it lowers their blood pressure and they live longer on average. This is the truth.

So, what's going on?

We know that God does not have a physical presence. We might have a feeling about God's presence, but God does not have a presence like this page in the front of you. God does not involve Himself in our daily lives. He does not answer prayers. Yet, while the whole issue does not make sense, most people believe in God as a concrete entity.

Most people are fearful just talking about the Almighty and His existence. They have a deep feeling that He *does* exist, that He is everywhere and that He is watching us. However, if you told them that you spoke with God, they would look at you like you were crazy. If you insisted that you have daily contact with the Almighty you might be sent to a mental institute. You can see the paradox: God exists, but He does not talk to average human beings unless they are leaders who have followers.

Why is it that different people always say, "God said this" and "God meant that" or "God promised this," while God never revealed Himself and explained what He really wants? We get our ideas about God through other people instead of from Him directly, and everybody sees things differently.

So, where did all this come from, where did we come from, where are we going, who are we, why do we have the urge and the need to believe in a higher power, where did the power of religion come from and where did God come from?

What's going on?

For these answers we have to look at the whole picture from the beginning, because everything is connected. Every single thing is the result of a chain reaction. Everything has a reason, a cause and an effect.

2. A True Story for the Believers

A few years ago in late summer, a couple days before September 11 of 2003, my friend and I traveled from Philadelphia, PA to Huntington, WV to visit some friends. It was a beautiful eight hour drive that didn't seem so long. It was a clear, sunny day and the scenery of West Virginia's tree covered mountains was exhilarating. Nature felt very close. The different shades of green were easy on the eye and calming to the mind. As we drove into the evening we saw beautiful endless sunsets. The sun kissed the top of the hills and the mountains as we headed west, until it finely disappeared in the western horizon.

We arrived in Huntington on the evening of an average Tuesday. The town was quiet and without traffic as we drove to our friend's house. After a warm welcome, we decided to go together to our favorite restaurant, where some of my friends had alcoholic beverages while I drank water. (I like to drink a lot of water before a meal. My body does not tolerate alcohol and it does not make me feel good.) We had a great time talking about everything, the food was excellent as usual and time passed by so quickly that we left the restaurant when the place was about to close.

After we all took a stroll downtown, my driving companion and I drove the rest of our friends home and headed to our motel rooms. On the way we saw the beautiful night sky. It was a clear dry night and the stars looked so close and numerous. We decided to stop at the top of a hill where there was less light pollution to take a good look at the heavens. We turned off the engine and the headlights and got out of the car.

After a few minutes I saw a glow of light in the nearby bushes, about eighteen yards away. When I told my friend about it he said he didn't see any glow of light in the bushes. I decided to get closer, to about three feet from the glow of the light, which was like a ball nine feet high and by now very bright. The bush leaves seemed to change colors every few seconds and the whole area was lit up like the glory of a morning sunrise, with a pleasant heat coming out of the glow. I knew it was still nighttime and there was no way I could see the reflection of the sun this early. I looked at my watch; it was midnight exactly, but the second hand on my watch did not move at all. I looked at my friend. He stood next to the car motionless, looking at the sky.

Nothing was moving and it was completely silent. It looked like time was standing still. Then, I heard a deep and very clear voice coming out of the glow and calling my name: "Elli...

Elli…," which means "my God" in Hebrew, and I said, "Yes, who is it?"

The voice replied, "I am God. I am the Creator. I want you to deliver my message."

I said, "Oh my God, I'm at your service," as the voice continued:

"Human beings are killing other human beings. They love to play with guns because of sexual deprivation. They also like to hit tall buildings because of sexual depravity, killing innocent people and cutting the joy of life short. Sometimes they sacrifice animals and say they do it for me, but I find this all very appalling. I do not need anybody to protect or kill for me. I am capable of killing by myself. I do not need anybody praying for me or calling me the Great One. I am healthy, immortal and have no competition up here, but I made some mistakes. I created a cruel and unforgiving nature.

"I want you to put the word out that I have nothing to do with those human beings. Most of them are stupid all of the time, because they cannot recognize a pattern, and the rest are stupid most of the time. They all kill and consume other life forms for their own survival, and some say they do it for me. I want you to put the blame on nature for the killing. There is a genetic code that nature has embedded in every human being and animal to keep them in the same order and to keep their own identity, to get rid of the competition because of limited natural resources and to please the leader by following him and sacrificing to him."

I asked, "Why me God? Why not my friend?"

He replied, "Moses said the same thing when I revealed myself to him in his imagination on Mount Sinai. He did a pretty good job even though he was slow and tired from climbing to the top of Mount Sinai and having very little food and drink for forty days and nights. Your friend is blessed already. He has a nice family, good health and a great life. You are a deeply religious Atheist. You wish for my existence every day and I like that. You are also fearless and have the inside knowledge of nature."

To this I responded, "I still have fears, like the fear of heights, and sometimes I get upset and eat like a pig."

God said, "It is normal to have fears. It is genetically embedded in you, to prevent you from getting into dangerous situations that can hurt or kill you, or to make you act very carefully if you do get into a dangerous situation. Eating like a pig is also normal because human beings are genetically closely related to the pig family. Therefore, you should avoid eating pork because it is borderline cannibalism."

"But I am still hard at tongue and a slow thinker. How can I put your word out?"

God replied, "You can write."

"Write?" I said, "The last time I wrote was over thirty years ago in high school and maybe a few love letters later on. Besides, I do not know what to write about."

"First you write what I'm telling you now," the Voice replied, "and then you write about the inside of nature from the beginning. It will be natural for you to write and the information will be revealed to you as you write. Everything will be original and you will have the insight of nature like no other man. Perhaps the numbers and the timing won't be accurate, since there is no absolute. Everything fluctuates in time and time itself is not absolute. But the principle will be right."

He continued, "I want you to tell everybody that I do not listen to any wishes or prayers. When two armies go to battle with one another and pray to me for victory by killing the people on the other side, it does not make any sense. Where is the brain in that?

"There is no mercy; I do not know the meaning of that word. Mercy on whom? If you go into a pool of water where there is a big hungry alligator, the strong one will win. I won't come down to save either of you. I let nature take its course. I do not intervene in anything. It is like I do not exist. You are all alone, and your destiny is in your hands."

His voice moved closer, "I really exist only in your imagination, which is the fifth dimension, the extra dimension people add to the four dimensions of this universe. With it they can bypass all of nature's laws. You can be anything you want to be, anywhere and anytime, as long as you have imagination, as long as you have a brain. Therefore, there is a god in every human brain that wishes for me in their imagination, but when the brain is dead, imagination is no more and I **do not exist**."

He then explained, "Animals may have a brain, but they do not have imagination like you do and therefore do not have a god. There is evil in your body that needs to consume in order to survive by killing other life forms. The god in you has to control the evil in you. I know it is a tough game for you because you are a product of nature and there is no way for you to disconnect your imagination from your brain or your brain from your body.

"It is as if you like to watch the pictures that television produces, but do not like to watch the television box itself, so you use mirrors to see the television images that the mirror reflects from the television box. Now, you try to get rid of the television box and still see the images in the mirror, but you know it is impossible without the television. The television produces the images, and without it, there are no images. Your brain is like the television; it has to be made of the right stuff, to have the right connections, and needs energy and information to produce images, or rather, imagination. In other words, if there is no brain, there is no imagination, and therefore, there is nothing.

"I have no body. I am all imagination. So I do not kill or consume. I am all good and I live forever, but I'm all alone here. I do not even have a telephone. There are no other gods to talk to and it gets boring. So, I play games all the time. I created this universe by exploding a small single graviton into many pieces, more than one with half a million zeros behind it, and there is no way for me to follow all those small and huge numbers of gravitons and the interactions between them. It is like a game of dice for me, a game of dice in the dark with infinite possibilities and an outcome I cannot see."

He paused briefly to let this sink in before continuing, "There is uncertainty built into nature. There is no way to gather all information all of the time to allow the prediction of future events. It is not fun to know everything because it takes away the excitement and the surprise in life. Do you think the Wright brothers could foresee how far their invention of the airplane would develop in just one hundred years? In their wildest dreams do you think they could imagine how far and fast airplanes would one day fly, carrying millions of people in the air at any moment of the day?"

His voice became softer and sadder, "But I knew that human beings would use the airplanes to drop bombs on other human beings. Life is so precious but so short, and you have only one life to live." The voice became more intense, "Only the human brain can conceive the universe, and in this way every human being is like the whole universe.

"Everything that has a beginning has to have an end. Everything has a lifespan. Everything changes forms and everything is interchangeable. Everything in this universe is alive except for black holes. Atoms are alive because they are always in motion. They communicate by way of photons and are attracted to other atoms by way of electrons and gravity. Like everything in nature, they also die and are reborn. Your body is alive. It is made from live atoms.

"A single bullet, or even invisible bacteria with no brain, can stop a human life. People have to look out for one another and not do to other life forms what they would not like done to them. Try to have the best life possible, as you see it. You are a product of the nature of this universe. Nature controls you and you control nature to a limit, depending on your energy and knowledge. You are programmed by your genes to survive and to pass on those genes to the next generation… like a carbon robot."

A brief pause again, "You are all connected by gravity, matter-energy and space-time. Everything is the result of a chain reaction and obedience to the laws of nature. You are made from matter that was born in this universe and will die with it.

"Some of this matter was put together by your genes to make your body, so you cannot be more than you were designed to be. Certainly you could be less, but never more. So enjoy life! The joy of life is holy.

"While it can be hard to stay alive and to live the way you want to, it is easy to die and to be destructive toward nature and other humans. If you start terror, you will live and die in terror and violence. What goes around, comes around. This is the law of nature. Just like the hands of time, the end will resemble the beginning.

"There is no Heaven and no Hell. You can create Heaven or Hell on Earth; it is up to you. Respect your parents and elderly; you will be like them sooner than you think. Treat your fellow human beings as equals. Women deserve more respect; they are your mothers and the mothers of your children and they have a harder role in life. Teach your children how nature works and to respect everything; you will have a better future if you do.

Be polite and helpful to neighbors and strangers alike; you will be happier and will live longer. The meaning of life is survival and the best way to survive is through harmony and love.

"If you do not enjoy life or if there is no joy in the future, life should be eliminated to make room for other lives. Nature is limited and we have to continue the cycle of life and death because life follows death and death always follows life."

I was all ears and clinging to every word as He spoke, "A birth of new life is joyous, and therefore is holy. To give, to help, to love or to make somebody happy is joyous and therefore is holy. Christmas is the holiest holiday. It is a giving holiday and a celebration of the birthday of my only human son. ...I play games; she was only a twelve year young virgin. It is true that planet Earth is the only place in the universe that has real and not imaginary virgins..."

He continued, "Prayers are not holy. I know they make you feel better because you think maybe a supernatural force will intervene to give help when you need it, but that is only wishful thinking. If you want things to happen, it can only be done through work and energy. You cannot create something out of nothing. You can create something out of something else and only in the frame of the laws and possibilities of this universe.

"I do not listen to any prayers and I do not do anything about them. Therefore, houses of prayer are not holy and there is nothing special about them. They are all made from the same materials with which you build slaughterhouses and bathrooms. They can be destroyed in time by fire, bombs or natural forces. People die in them just like in any other place. You can live without houses of prayer and you might get rid of slaughterhouses in the future if you use your brains correctly, by using genetic technology to grow meat in factories, but you will always need bathrooms."

He paused briefly, "There is nothing holy about marriage. It is only a contract with witnesses that can be broken for many different reasons and does not guarantee your happiness.

"There is nothing holy about so-called 'holy men'. They are all average human beings with physical needs and they all have bodies that consume other life forms. They are all slaves to the urges and pressures of their genes and they all obey the laws of nature. They cannot come back to life after they die and they cannot perform miracles or supernatural events. All they can do is give you half the truth or complete lies. They create illusions surrounding the unexplainable things in life, playing with your fifth dimension.

"The preaching's of these holy men are only good for social gatherings and entertainment at best. If they do preach, they should always encourage love and understanding and never spread hate. The world could be a better place if all those religious preachers became agriculturalists, scientists or comedians. Earth would be a happy place and so would I.

"Humans are put together by nature against tremendous odds. Life is a gift; it is short and unusual. Death is the natural state; it is infinite and it is always waiting for you, so do not hurry to die. Life is a challenge. I help only those who help themselves, and if you truly can help yourself, you do not need me anyway and it is as though I do not exist.

"If any person calls to cut the joy of life short, not for survival, but for religious or superiority beliefs, you should eliminate them as early as possible, before they grow too powerful and create more harm. This is just like a cancer or a tumor that is easier to remove in its early stages with a knife or other earthly forces.

"Although the universe is full of paradox and things that have more than one meaning or explanation, your inability to explain certain things doesn't mean that gods or creators must exist.

"It is funny that so-called "holy wars" are always fought with manmade devices like stones, knives and weapons, all of which have nothing holy about them. The same is true about "holy" places that are no different than other places.

"You can eliminate evil forces by greater forces or superior knowledge. This is the law of nature and I do not intervene. You have to use your big brain, which sets you apart from the rest of the animal kingdom, to have a better life. You are blessed with a big brain. Its main purpose is the ability to see farther into the future in order to calculate the odds in your favor and to satisfy and control the pressures and urges that come from the genes in your body and mind.

"Time only goes forward and you have to flow with it. Laws that were created two thousand years ago, or even one hundred years ago, do not work as well in today's world. Things change with time and laws have to change to reflect changing conditions.

"There is nothing absolute in nature. Does Saturday come before Sunday? Of course. Does Sunday come before Saturday? Of course. Nature is pure. There is nothing ugly, beautiful, bad, good, big, small, right or wrong in it. It is just the way it is; everything is relative and only your brain sees it differently with its ability to compare and to imagine.

"If you try hard enough, nature will reveal all of its secrets for you to harness and enjoy, but you won't be able to go back in time or to change any of the basic laws of nature. You won't be able to change the fate of your universe, because you are part of it.

"You came from nothing and you are going back to nothing, but in between you will make some waves that affect other people. Some people will make more waves than others, for the good or the bad. Life is a game, and the name of the game is "Real". As long as you live, everything is real. There are a lot of games within the real game of life. Enjoy the games and play them right. Give more joy to more people more times, because when you are dead you are out of the game, and there is nothing. Even when you are alive you live in a paradox: you are nothing and yet you are everything, you are smart and yet stupid, you are all the same and yet you are all different.

"I bless you and I bless this book, which can outlast all other religious bibles to stay with human beings as long as they remain human beings, maybe for billions of years to come. This is the only holy book that can bring true peace to mankind, as much as peace is possible, since it shows the limitations of humanity and the limitations of your small universe."

With these words, the glow of the light disappeared from the bush. I felt very sad and lonely. I had wanted it to last forever. Suddenly, I felt a strong urge to write God's words. I looked up to the heavens full of glowing and twinkling stars, and then I looked down at my watch; the phosphorous in it was glowing from the absorption of the glowing bush. The time was midnight exactly, but now the second hand was moving. I looked at my friend. He was still in the same position standing next to our car, looking at the heavens. When I walked toward him he looked at me and said, "What a beautiful night sky. Did you see anything in the bushes?"

"I have seen God's glory," I said, "Let's go to the motel."

We watched the beautiful night sky in silence for a few more minutes, then got into the car. My friend started the engine, turned on the lights, looked at me closely and said, "You look sad. Is everything ok?"

"I don't know," I said.

"It looks like your ears got bigger and you lost more hair," he said.

"Yes, but I gained some knowledge and I just became a born-again Atheist," I replied.

"A born-again Atheist? What does that mean?"

"I know that God does not exist."

"And how do you know that God does not exist?"

"God Himself told me so," I said sadly.

He asked with a smile, "God himself told you so? Isn't that an oxymoron?"

"I think it is an oxymoron, but who am I to dispute God's own words?"

"What else did God tell you?" asked my friend in a sneaky way, checking my sanity.

"He told me the truth," I answered right away.

"And what is the truth?"

"Can you handle the truth?"

"I think so," he said with fear in his voice.

"God said that we have only one life to live, we came from nothing and we are going back to nothing..."

I came back to Huntington many times after that night to look for that spot on the hill, but it was never to be found.

3. Genesis

Before our universe came into existence, there was, as there still is, and will forever be an infinite universe, a size-less universe with no boundaries. It is a timeless universe with no beginning and therefore no end. This infinite universe is actually a Black Hole, a realm of emptiness (no space) that does not transmit any information and is pitch black with no light to see anywhere.

This infinite universe contains an infinite number of single-dimension objects of different shapes and sizes with no connections or influence of any kind between them. These objects come into existence from nowhere and go back to nowhere in an unspecified amount of time. All these objects are dead, meaning they don't have any internal motion. They are floating motionlessly in the unlimited emptiness of the infinite universe.

About fifteen billion years ago there was a random collision between two objects of similar size and dimension but of opposite charge. One had a positive charge (male) and one had a negative charge (female). The collision awoke God from his eighteen zillion year nap to put order and life into the newborn universe in a simple way, by creating many dimensions of space in addition to one dimension of motion. As you will see later on, this gave us the four basic laws of our universe: gravity, matter-energy, space-time and speed limit (the constant of the speed of light). A new universe was born, an infinitely small and insignificant universe compared to the infinite universe around us.

Our universe is one of an infinite number of universes with different shapes, sizes, dimensions, lifespans and directions of evolution. It floats motionlessly (meaning there is nothing around us, of which we are aware, to which we can relate our speed) in the unlimited emptiness of the infinite universe. There is no connection or influence of any kind between the different living (meaning they have internal motion) universes.

The evolution of our universe started about fifteen billion years ago in what we call the Big Bang in many steps that continue to this day, and it will end in about fifteen billion years from now, giving our universe a lifespan of about thirty billion years. There were three main stages of the Big Bang that led to the creation of our universe in its present form.

The first stage of the Big Bang was dark and silent, and was due to a random collision between two similar objects of single dimension, but of opposite charge. That collision created an explosion, a splitting of the matter-energy belonging to these two objects. The explosion shot out many trillions and trillions of single gravitons (which are positive attraction units) to create a ball-shape sixty billion light-years in diameter. All of these single gravitons were held away from one another by lines of repulsive energy. This defined our universal space, its shape and borders, like a universe-wide-web. It all happened in no time; time wasn't born yet.

Now the second stage of the Big Bang, while everything was still dark and silent, some of the single gravitons that were denser in the space of the initial collision (which was about six light-years in diameter) paired together (by collision and attraction) with other single gravitons (with no distance between the two gravitons) to create the building blocks of matter in the form of photons. Light was born, shooting outward and away from the central point of the Big Bang in all directions at the speed of light. Time was also born and began to tick forward as these photons, because of their motion, changed relative position to one another and also relative to our universe space. God said it was a beautiful light show that lasted six long days without nights (about six minutes of our time).

Within these six days of light the third stage of the Big Bang occurred through shockwaves and sound waves. Matter and antimatter were created by the many photons colliding among themselves to form the very small subatomic particles, mostly the neutrinos that are made from at least two building blocks of matter with a distance between the two gravitons of which each building block is made. The two gravitons that create a single block of matter are connected together by one line of energy and are spinning around each other. The direction of the spin of the building blocks of matter also creates the same directional spin in the subatomic particles they create by way of grouping with other building blocks of matter. In turn, all these subatomic particles were spinning either clockwise or counterclockwise to counter their own gravitational force and shooting outward and away from the point of the Big Bang. Matter and antimatter were born almost in equal quantity, with an advantage of about one percent for regular matter which spins clockwise.

Within a few days after the Big Bang most of those newly created subatomic particles became larger and more complex as they collided with other subatomic particles and photons in this hot and particle dense environment. Because those particles were unstable they were easily attracted to other unstable particles until they formed stable particles: electrons and protons.

The attraction of the electron to the proton created the hydrogen atom and these simple hydrogen atoms were born in huge numbers.

In the event of a collision between matter and antimatter, the two would annihilate one another to form photons.

Just like regular matter, antimatter has four dimensions and obeys the same laws of physics, including the force of gravity. The difference between matter and antimatter is simply the direction of the spin of their subatomic particles, all the way down to the spin of their building blocks. Matter spins clockwise and antimatter spins counterclockwise, giving them an opposite charge. In the atoms of antimatter the electron will be positive and the proton negative and therefore there is always a strong electrical attraction between matter and antimatter.

All matter and antimatter atoms (including subatomic particles) spin to offset their own gravitational pull, and by doing so produce magnetic fields that give them negative and positive charges at their poles. Any contact between matter and antimatter is always a high speed collision. Since the subatomic particles of matter and antimatter spin in different directions at speeds approaching the speed of light, they therefore annihilate each other to become photons, flashes of light.

On the other hand each photon is made from one pair of gravitons with no space between the two gravitons and therefore this pair of graviton does not spin in either direction which makes the photons unique: they are not in the matter-antimatter category. Photons are the only particles that can interact with a matter object equally as with an antimatter object creating the same end result. As you see all matter, antimatter and photons are made from pairs of gravitons. The difference between them is only the distance between the two gravitons in the pair and the direction of the spin of the pair. So throughout this book I've written "matter" or "matter object" but it might as well include antimatter or photons since they are all made from the building blocks of matter: the pair of gravitons.

The constant attraction and collision of matter and antimatter continued until most of the antimatter disappeared due to its disadvantage of one percent in quantity. This formed gaps in the uniform spread of matter from the point of the Big Bang.

A few million years after the Big Bang, these gaps in the cloud of matter (which by that time consisted mostly of hydrogen atoms) grew in size. At the same time our matter universe grew in size and slowly cooled down to form many billions of different galaxies.

The giant clouds that cooled down lost most of their energy by losing heat, and started to spin as a means to balance the gravitational force of their own mass. As they cooled down further from the expansion of our universe matter, their centers collapsed in on themselves. They didn't have enough energy in the form of heat and of spin motion to counter their own gravitational force, and therefore formed black holes. A black hole is just as its name implies: a hole in our universe space.

Our universe space is the huge number of single gravitons held apart from each other by lines of negative (repulsive) energy which act as a web that fills and defines the shape and size of our universe space. This web controls the speed limit and force of gravity by way of the numbers and the motion of the single gravitons.

All matter, by way of the presence of its mass in our universe space, creates weaker repelling forces between single gravitons that carry the matter and the single gravitons nearby, and the single gravitons nearby then move into the weak energy spaces, creating forces of gravity in the form of the movement of those single gravitons that carry mass toward the centers of matter.

Only energy in the form of spin or motion of matter can offset the force of gravity. That is the reason why all matter structures are in motion, including suns, planets, atoms and subatomic particles. The more concentrated the matter, the faster the spin or the motion.

In the case of a black hole a huge amount of matter is concentrated in a small place and the force of gravity is so great that it overcomes the mass energy. The matter loses all of its space and therefore collapses into itself. It becomes a different dimension and creates a hole in our universe space equal to its mass.

The more mass falls into the black hole, the bigger the hole becomes and the stronger its gravitational force becomes to equal whatever amount of gravitational force the matter had. The matter inside the black hole becomes a different dimension, losing its identity as matter, and leaves our universe at a speed above the speed of light without a trace. No matter, not even photons with a velocity of 186,000 miles per second, can escape from the black hole surface. That's why we call it a black hole: no light can come out of it.

So, in the early universe all the big galaxies formed at least one giant black hole in their center where most of their mass was concentrated. Surrounding the black hole of each galaxy, but many billions of miles away from it, was enough matter in the form of gas clouds made mostly from hydrogen atoms to create millions of giant stars.

This process happened every time that the pocket of a gas cloud collapsed into itself to form a giant sun ten to one hundred times the mass of our Sun with such a tremendous amount of gravitational force as to squeeze the matter inside these suns, which was made mostly of hydrogen atoms that were fused together to create heavier atoms mostly of helium.

The byproduct of the fusion was energy in the form of photons that shot outward at a temperature of millions of degrees centigrade. This created motion and outward pressure inside the giant star, which offset the further squeezing force of its own gravity.

The bigger the star, the faster it burns. Blue Giants have a lifespan of up to a billion years; they burn very hot and shoot out photons, mostly in a short wavelength. Stars depend on their critical mass, which can be over one hundred times the mass of our own Sun. A star, or a pocket of matter, can collapse into itself without igniting or becoming a shining star. This way it becomes a black hole directly because of its own enormous gravity force.

When a star is ten to one hundred times the mass of our Sun it will burn fuel by fusing its hydrogen atoms at a faster rate, in between a few million and one billion years depending on its mass. Once the star has exhausted its hydrogen it will collapse in on itself to create more pressure and heat, which causes the helium atoms inside of its core to fuse into heavier atoms.

Once the core becomes mostly made of iron atoms and heavier elements, the fusion stops. Since those atoms need more energy than they can produce by fusion, the star collapses in on itself. Its outer layers burst in a Supernova explosion, with shockwaves traveling away many billions of miles at about half the speed of light, shooting out particles of matter no bigger than stripped atoms that can withstand the explosion force. The sun's core collapses in on itself and depending on its left over mass quantity, it either becomes a black hole, because it does not have enough energy to withstand its own gravity force, or it becomes a Neutron Star.

A neutron star is the left over core of a large burned-out star. The left over core is made mostly of iron and heavier atoms that are held together by their own tremendous gravity force. Their gravity packs them together very tightly. Only the spin and the movement of their subatomic particles gives them enough energy to withstand this gravity force and to keep them from collapsing further down, in which case they would become a black hole.

In the case of the neutron star, the electrons that circle the nucleus (the protons and the neutrons) collapse in on their nucleus because of the tremendous gravity force. The electrons, which have a negative charge, fall on the protons, which have a positive charge, and they cancel each other to become neutrons. In this way a neutron star is born.

A neutron star is only a few miles in diameter, but packed with a lot of mass. One cubic inch of its mass will weigh several million tons on the surface of the star. The reason is that the neutron star is made from neutrons only, not from atoms and molecules. Atoms have a huge distance between them. For example, if the atom nucleus is a mile wide, the nearest other atom nucleus will be about a hundred thousand miles away. In the case of the neutron star, this space is eliminated and the neutron star's density becomes equal to any atom's nucleus mass density.

The neutron star spins very fast, at close to half the speed of light, and also rotates several times a second. The fast spin gives it energy to withstand a further collapse, in which case it would become a black hole. The fast spin of the neutron star also creates a huge magnetic field at its axis. All matter creates magnetic fields (photons) because of its spin. In the case of a neutron star, which is tightly packed with very heavy matter spinning very fast, the magnetic field is much stronger. It is the strongest magnetic field in our universe, with enough energy to heat the gas and dust cloud around it for many millions of miles. This heat produces photons that because of the neutron star's rotation can be seen from Earth as pulsating lights. That's how it got its second name: Pulsar Star.

Because of its super density and weight, the neutron star creates a tremendous amount of gravity force at its surface (second only to a black hole's gravity force) that will suck in any matter object nearby that doesn't have enough energy to withstand the neutron star's gravity force. Any additional matter falling on the neutron star will make the star bigger and heavier, giving it a stronger gravity force. To compensate for the stronger gravity force, the star will spin more quickly to therefore produce a stronger magnetic field.

As it sucks in matter from its surroundings the neutron star can get bigger and heavier up to the point of critical mass. Once it reaches that threshold point, its own gravity force will overcome its mass energy and the neutron star will collapse in on itself to become a black hole. As time goes forward in a scale of many millions to a few billion years, those neutron stars that don't receive any matter will become smaller and weaker by losing some of their mass energy in the form of photons.

So, in the early stage of our universe, about a billion years after the Big Bang, the condition of the galaxies made them ripe for the birth of at least one giant black hole in the center along with a few smaller ones around it and many millions of blue giant stars. They shone for many millions of years before they exploded in a supernova, leaving either black holes or neutron stars behind. The exploding stars shot out a huge amount of matter in all directions through shockwaves of many billions of miles. They enriched the gas clouds composed mostly of hydrogen (of which galaxies are made) with heavier atoms in nature.

In the central and outer reaches of the galaxies there was enough gas and material to produce many billions of smaller suns, each about the size of our Sun, which could shine for many billions of years. After a few billion years, depending on their mass, the heavier suns burned faster than the lighter ones. Most of the hydrogen in each sun's core was fused to make heavier elements mostly of helium. The sun would then collapse from its own gravity force, creating more pressure and heat in its core to fuse the helium atoms together into heavier elements.

At that time the core of each sun swelled, and these suns became Red Giants, many times the size of the original suns. They remained this way for many thousands of years until their helium burned out. At that stage, the sun would collapse in on itself because it did not have the energy to offset its own gravity force.

The collapse of the sun in a few seconds creates a tremendous amount of pressure, and the sun explodes in a nova explosion, shedding its outer layer in a shockwave many millions of miles outward in all directions and shooting heavier atom elements into the galaxy space. The left over core, made mostly out of iron and heavier elements, shrinks to a few thousand miles in diameter, much smaller than the original sun that was up to a million miles in diameter. That left over core becomes a White Dwarf.

A white dwarf is a ball of matter made mostly from iron and the atoms of heavier elements that are held together by their own tremendous gravity force. The gravity force packs them so tightly that only the resistance of the electrons that spin very fast around the nucleus prevents the star from further collapse, in which case it becomes a neutron star or a black hole. This happens simply because it does not have enough mass and therefore gravity force to squeeze it further. The energy of its spinning as a whole along with the spinning of its subatomic particles is enough to offset its own gravity force inward.

However, the gravity force is strong enough to squeeze the electrons to a lower orbit, and by doing so it creates a ball of matter that is very tight and heavy. A cubic inch of this matter weighs over one hundred tons on the surface of the white dwarf.

The white dwarf spins very fast. It is not as fast as a neutron star, but just fast enough to give it energy to offset its own gravity force, which tries to squeeze it further. The white dwarf stays hot enough to produce photons for many millions of years by the collision of its own subatomic particles among themselves. That's how it got its name: white dwarf.

Over a period of millions of years, if the white dwarf does not receive any additional matter from its surroundings, it cools down because it shoots out more photons then it receives. It then becomes a Brown Dwarf, which is cooler, smaller and produces less photons. After a few million years, the brown dwarf becomes a Black Dwarf, which is a little smaller and much colder than the brown dwarf. It is only a couple of degrees above its surrounding temperature, which could be only about three degrees above absolute zero.

The galaxy also produces many billions of gaseous planets, giant ones many times the size of our own planet and some approaching about half the size of our Sun. They are made mostly of hydrogen and helium. Their gravity forces aren't strong enough to fuse their hydrogen cores and they don't burn as suns.

About nine billion years after the Big Bang there were enough heavier atoms in the galaxy space to create billions of smaller planets about the size of Earth and small rocks and dust to create rocky planets like our Earth. The heavier atoms are the end product of dying suns that fused their hydrogen atoms.

After a sun dies, it spreads most of its material through shockwaves that penetrate deep into the galaxy space to create more suns and all kinds of planets. Lighter planets always circle heavier planets in their vicinity to become moons or satellites. The big planets always circle heavier objects like white and brown dwarfs, suns, neutron stars and black holes. Black holes never circle any other object; they just move as the fabric of our universe moves.

All the matter in our universe is expanding with the billions of galaxies it formed, outward and away from the point of the Big Bang, just like dots in an inflated balloon.

This process of creating black holes, neutron stars, suns and planets, along with creating heavier atoms out of the mostly hydrogen cloud of which the galaxy is made, was happening more or less throughout the billions of galaxies in our universe, with its many different solar systems and different combinations of suns together with their planets, circling bigger suns, or suns circling around black holes.

There were many collisions of big objects such as planets, white, brown and black dwarfs, suns and even galaxies in the first nine billion years. As the matter-universe expanded and the distance became greater, there were less and less collisions.

About nine billion years after the Big Bang, there were enough diverse atoms to form a huge number of different molecules. The different and diverse atoms collided among themselves with the help of shockwaves created from the explosion of Death Stars or burned out suns. The dust and gas clouds compressed, clearing the way for chemical reactions among the many particles of which they are made. Diverse atoms attracted to one another to form many different molecules, mostly water and iron oxide, among others.

As time moved forward in a process that took millions of years, those different small molecules interacted among themselves. With the help of shockwaves and cosmic rays, the byproducts of exploding suns, they formed many bigger and more complex molecules. One of these was amino acid, the building block of organic life on Earth.

Some of those small and large molecules grouped together to form interstellar dust, again with the help of shockwaves, gravity force and collision, to form bigger objects like rocks that held together by their chemical reactions until they became large enough to have gravity fields strong enough to attract other small rocks and dust. These became comets.

Asteroids are mostly a product of the collisions between planets and left over suns, white, brown and black dwarfs, that sometimes leave large fragments hundreds of miles in diameter. Their nuclei are made mostly of iron and other heavy metals. The collisions and gravity forces of all those small and large objects created bigger objects, like small rocky planets. Our planet Earth is one of them.

4. The Laws of Nature

We must have laws to have an order. Without laws there would be chaos and we wouldn't be here to ask any questions.

The basic four laws of our universe are gravity, matter-energy, space-time and the constant of the speed of light.

These basic laws started with the birth of our universe, the Big Bang. They are connected and correspond to the four-dimensional universe in which we live, the three dimensions of space and the one dimension of motion.

The motion of gravity always pushes matter inward toward its center. The motion of energy always pushes matter outward away from its center. The motion of time is connected to the motion of matter and only goes forward, regardless of the direction of the motion of matter, which can go inward or outward.

As time went forward, matter became more complex, creating unique structures. Pairs of gravitons, which are the building blocks of all matter, grouped with other pairs of gravitons in different numbers and shapes (dimensions) to create all the different and diverse subatomic particles. The diverse subatomic particles grouped together to create the simple atom of hydrogen. The simple hydrogen atoms fused together under pressure and collision to produce the many different atoms. In turn, over a period of billions of years, some of those diverse atoms grouped together to create the many diverse and unique molecules in our universe. They also created our bodies.

At this point in time the basic four laws of our universe created many new laws; as matter structures interacted among themselves and became multidimensional, it gave rise to many complex laws of physics. These laws are temporary and they depend on location, quantity and the state of matter, unlike the basic four laws of nature that started in the Big Bang and will last until the end of our universe (about fifteen billion years from now).

All these laws are abstract, but nevertheless, they are real. They put an order to the positions of the huge number of matter particles of which our universe is made. By learning these laws we can see the pattern; we can read past and future events that are normally sealed.

We as humans create our own laws, since we do not live alone. We share Earth's space with plants, animals and other humans, and to have an order we must have laws.

We have governmental laws to keep order among people, to enforce the sharing and protection of space and energy, as well as to help the needy, because we all need help at one time or another. We have laws to enforce education and laws to collect taxes. We have laws that involve our motion, be it through air, sea or land. Otherwise, we wouldn't be able to survive.

We also have religious laws, some of which supposedly came from God, but were written by people like you and me. These religious laws may reflect the religious beliefs of the majority of people, or of the most powerful people, in a particular place. All these laws are temporary and they are good (or maybe bad) for their own time. In other words, laws have to be changed to reflect changing conditions and they are different in different locations.

We are inherently law-abiding, at least to some laws. Otherwise, we would not survive as a group, or even as individuals. For example, it is a law of nature that we all have to consume food for energy and to build our bodies.

5. Gravity

We start with the law of gravity, the mother of all the laws of nature, the connection of all the matter in our universe to the fabric of our universe (our universe space) and the basic force that connects all matter structures in this universe to one another.

The belief that matter objects or black holes create gravity or the pulling of matter objects to their surface is false. The fabric of our universe creates the gravity phenomena in reaction to the presence of matter objects or to the emptiness of black holes. Just like a virus in the human body doesn't create fever. The body immune system creates the fever in reaction to the virus present in the body.

To understand gravity we have to go back to the moment of the Big Bang, which was the splitting of matter of a different dimension into a huge number of single gravitons that burst outwards carried by negative energy lines to occupy an emptiness of approximately thirty billion light-years in all directions away from the point of the Big Bang. This defined the borders of our universe, its space and its functions.

These single gravitons created the fabric of our universe and imposed speed limit, gravity and time on all matter structures, including the photons in our universe space, just by way of their numbers, their motion and the space between them. Single gravitons are positive (attraction) energy units that are attracted to other gravitons nearby, but in the gaps between them there are negative (repulsive) energy lines that hold the single gravitons in their place. At the point and the time of the Big Bang, when these single gravitons were denser, some of them paired up by collision and attraction with other single gravitons to form the building blocks of matter.

So basically, all photons, matter and anti-matter are made from pairs of gravitons which were a different dimension that changed form to become the matter we know, the matter of which we are made, just by pairing up. By pairing up, the two gravitons that create a single building block of matter occupy a smaller space between them relative to the space between the single gravitons that are part of the fabric of our universe.

Gravitons are positive energy units, only a certain number of gravitons per given amount of space can occupy that amount of space. When pairs of gravitons are added to a given amount of space it creates imbalance between the positive and the negative forces in the space between the single gravitons: the negative force gets weaker and a gravitational field is created. The building blocks of matter, due to their presence on single gravitons, hold more gravitons per given amount of space relative to the fabric of our universe, and so create weak gaps. This means they create a weaker repelling force (the negative energy line that repels single gravitons from each other becomes weaker) between the single gravitons that carry matter and the single gravitons nearby. The repelling force gets weaker and shorter for the duration that matter or photons appear on the single gravitons. The nearby single gravitons stream into these weak gaps and act like a wave carrying all matter into its central gravitational field.

So a gravitational field is a field of low repelling force (low negative energy) between the single gravitons. Therefore, central gravitational field would be the lowest point in that field. The flow of single gravitons toward the central gravitational point as a wave is constant, but they only flow if they carry matter, and when they do, for this duration their movement is the **force of gravity**.

Basically, all matter including antimatter and photons, creates weak gaps in the fabric of our universe since all photons, as well as all matter and antimatter are made of pairs of gravitons (positive energy units). The greater the amount of matter and the more concentrated it is, the more weak gaps it creates in the fabric of our universe. The closer the fabric of our universe to those matter objects or holes the weaker the negative force in the gaps between the single gravitons become, which leads to a stronger attraction force, all in accordance to Newton's laws of gravity.

So, the weaker the negative (repulsive) force in the fabric of our universe becomes the stronger the positive (attraction) force becomes. Meaning, matter objects will attract each other with a greater force and will get closer to each other (becoming denser). This includes the building blocks of matter: the pairs of gravitons. Under a gravitational field, the negative energy line that repels the two gravitons in each building block of matter from each other becomes weaker and shorter. The stronger the gravitational field, the shorter the space between the gravitons becomes, which leads to a denser, heavier matter object as it can be seen in the center of suns and planets. Neutron stars create the strongest gravitational field from any other matter object, second only to black holes and therefore the building blocks of matter would lose about 50% of space between the two gravitons and the matter will become very dense and heavy.

Approaching black holes where the negative force is very weak (almost zero) the building blocks of matter would lose all the space between the two gravitons in each pair (the negative energy line between the two gravitons will disappear) and become photons (a photon is just like the building blocks of matter. It is made from two gravitons, but with no space between them). Once the photons enter the black hole's emptiness they explode inwardly. Because black holes do not have any negative energy (negative energy counter balances the positive force) the positive (attraction) force that the gravitons are made of becomes infinitely strong and the gravitons (explode inwardly) collapse in on themselves to become infinitely small and dense objects, if any at all.

Now, let's go back to our universe. When matter objects move, the gravitational field they create by their presence moves with them and can be detected right away anywhere in our universe (I wonder if we can harness this phenomena for instant communication over really long distances). The weakening of the negative energy lines between the single gravitons due to the presence of mass or holes in our universe space happens right away, it doesn't have a speed limit. The fabric of our universe is subject to different laws of physics that may seem very strange to us. For example, let's take our solar system to see how gravity affects the motion and position of all the matter objects our solar system is comprised. Our sun is the heaviest object in our solar system and therefore it creates by its presence the biggest and the strongest gravitational field in our solar system. The lowest point in this gravitational field is always the center of the sun.

Therefore, our Sun is always situated in the center of our solar system (and not the Earth with its human population). All matter objects from planets, comets, asteroids and dust circle the sun. The closer these objects are to the sun, the shorter the time they need to circle the sun. So what will happen if the sun suddenly disappeared? All these matter objects that circle the sun will keep moving at the same speed but in straight lines even though they still can see the sun (they do not see the sun, they see only the photons that left the sun before it disappeared).

According to Einstein gravity is a result of matter-energy bending or curving space-time (his majesty doesn't explain what space is, what time is or how the actual curving of space occurs. He came to the correct answer of how gravity works blindfolded and without God whispering in his ear...Amazing!!!), and this curving of space happens or disappears at the speed of light. In other words, if our sun disappears Earth will keep in its orbit around the sun like nothing happened until it actually sees the sun disappear, since according to Einstein the curving of space has the same speed of photons (light). I think matter-energy curves space right away and if matter-energy disappears, space becomes flat right away with no time lag. This thing is very hard to prove since no matter-energy can disappear in our universe. Matter-energy can only change form from energy to matter or matter to energy (photons), or move its location with a speed up to the speed of light, or becomes a black hole.

So, if our sun becomes a black hole, this black hole will create the same gravitational field as the sun. Earth will circle this black hole at the same speed and distance as it did around the center point of the sun.

If our sun suddenly explodes and all the matter it's made of becomes photons all at once, it will create an expanding ball (sphere) that its envelope is made from photons moving outward at the speed of light. Earth orbits around the exploded sun will stay the same, until Earth actually sees the sun explode (about 8 minutes after the actual explosion) and then Earth's motion will be somewhat in a straight line. The gravitational effect of our sun on the nearest solar system, which is about 4 light-years away won't change as soon as our sun exploded but only about 4 years later when this solar system can actually see our sun's explosion because it takes about 4 years for the light to reach them.

So, according to the last two examples it seems as the gravitational wave has the same speed as the speed of light--*Not really*. When the sun exploded it turned all the matter it's made of into energy (photons) all at once creating a perfect round ball expanding outward at the speed of light. The combine mass of all these photons is the same as the combine mass as all the matter the sun was made of, creating the same gravitational field with its center point where it was before. So, as long as this sphere is in front of you the gravitational wave doesn't change, once the envelope of this sphere passes you, you will be inside the sphere and the gravitational wave will change directions and intensity relative to you.

The speed of gravity waves (single gravitons) varies. It is close to zero when objects of matter are distant, about ninety percent of the speed of light when single gravitons are near the surface of any atomic nucleus or neutron stars, and equal to the speed of light when they are near the surface of any variably sized hole (black hole) in the fabric of our universe.

Since the gravity waves are coming in from all directions, gravity force will squeeze any matter toward its center and the matter will become round.

When far away from any object that creates a gravitational field, matter structures always create a gravitational field of their own, as the single gravitons move towards matter. This will happen even during free fall or under conditions of weightlessness, because the presence of any matter creates shorter gaps in the fabric of our universe.

The repulsive energy lines between single gravitons hold the building blocks of matter apart, preventing them from exploding and changing dimension, since all matter is also made of positive (attraction) units of energy.

Because single gravitons are smaller and have less energy than the building blocks of matter (they are even smaller then photons), they are impossible to detect. Single gravitons belong to a different dimension and obey different laws of physics. It is only because of the effect of gravitational force and the speed limit of matter structures that we can know of their presence.

The force of gravity is very weak in comparison to magnetic force, but because magnetic force is a byproduct of the spin of matter structures (such as electrons, atoms and planets) that gives them the energy to counter their own gravitational force, magnetism is really a byproduct of the play between the forces of energy and gravity. Indeed, without gravity, there is no magnetism.

The force of gravity is extremely strong when generated by concentrated matter, such as a neutron star or even an atomic nucleus. The force of gravity weakens over distance but it never truly disappears, unlike the force generated by a magnetic field, which rapidly dissipates over short distances. Gravity, unlike magnetism, always attracts and never repels, both in the case of matter and antimatter.

There is no shield against gravitational force. The only thing that can act against gravity is energy, which opposes gravity like its mirror image. Energy direction is always outward, away from matter, whereas gravity direction is always inward, toward matter.

Here are a few examples:

When you stand on Earth's surface your weight is a result of deceleration: Earth's surface stops your motion downward. This motion is created because of Earth's gravitational force downward. On the other hand, your weight can also be seen as a result of your body's resistance to Earth's acceleration outward.

If you take a spaceship into deep Space, far away from the gravitational pull of any object, and use energy to accelerate this ship with a force of one "G" (the same force of gravity on Earth's surface) by firing its rocket engine, the people inside the spaceship will not feel any acceleration or movement. However, they will feel that they are under a gravitational force that pushes them to the floor, just as if they were standing on Earth's surface.

If we take one pound of matter on Earth and put it on a scale, the scale will show one pound because the gravitational force on Earth will push this pound of matter downward on the scale with a force equal to one pound. However, if we lift this one pound of matter to a certain height and then drop it on the scale, the scale will show two pounds. If you take a picture of this scale when it shows two pounds, you will think when looking at the picture that the pound of matter weighs two pounds. Therefore, the energy of matter in motion can be seen as a gravity force.

A practical use of this gravitational force is the increase of spaceship speed in the "slingshot method". In the slingshot method, a spaceship is allowed to fall toward a planet, and as it gains speed from the planet's gravitational pull, a small side rocket is fired, causing the spaceship to change direction and to maintain its new velocity.

Basically, all matter structures need either energy or gravity to accelerate, decelerate or change directions. All matter objects are always in motion. This is true for the smaller objects (the building blocks of matter, the pairs of gravitons, the subatomic particles and the atoms) as well as for the larger objects (planets, suns and galaxies). Matter has to be in motion all the time, which gives it the energy to offset the inward push of gravity. Without energy of motion, matter would collapse in on itself, lose all of its space and become a different dimension.

All bodies of matter create gravitational fields that attract other bodies of matter. The closer they are to each other, the stronger the attraction. The greater the mass of matter, the stronger the gravitational force it creates. This means that the lighter object will always circle the heavier object. The motion around the heavier object will give the lighter object the energy to resist being swallowed by the heavier object. Therefore, if we have a heavily concentrated object and a much lighter object nearby, the lighter object will circle the heavier object at a high speed to give it the energy to resist being swallowed by that much heavier object.

We see this in the case of an electron circling its atomic nucleus, which it does at a speed approaching the speed of light. We see this in our Solar System, where all the planets circle around the biggest and the heaviest object within it: our Sun. The closer a planet is to a sun, the faster it will circle around it, because the faster circular motion will give the planet more energy to counter the stronger gravitational force of the nearby sun.

All matter objects in our universe are spinning around themselves to counter their own gravitational force. The more mass they have, and the more tightly that matter is packed, the faster they will spin. The byproduct of the spin will be a magnetic field at the poles. One field will be negative and one positive depending on the direction of the spin, just as it happens in matter versus antimatter.

The force of gravity on Earth may seem weak to us. An average human can lift one pound of matter with no problem against the whole of Earth's gravitational pull. But don't forget that every pound on Earth weighs a pound because of Earth's gravitational pull, and that's a lot of pounds. You would need a tremendous amount of energy to lift all those pounds. If all the atomic bombs that humans possess exploded all at once, shooting matter debris into deep Space, the lost matter wouldn't be enough to lower Earth's gravitational pull.

Gravity is a destructive force, because you always need energy to counter it and yet there is no life that we know of without it. Gravity is the force that binds us together. We can walk on Earth because of it. Gravity is the force that holds matter-energy together, acting as a lid that prevents matter-energy from exploding. Just like Earth's atmosphere prevents liquid water from becoming gas, gravity prevents us matter-energy objects from exploding.

Without gravity, everything would tear apart in literally no time, as it happens in a black hole. Gravity is the first law that gives us life and will be the last one to push all matter behind the boundary of our universe into the black hole that is the emptiness around our universe. This will happen in about fifteen billion years from now.

We will expand on gravity in the next few chapters.

6. Matter-Energy

All the matter in our universe was created in just six days during the second stage of the Big Bang, which was the origin of all matter-energy in our universe. All this matter was made from the building blocks of matter, that is, from pairs of gravitons.

Some single gravitons, which were part of a different dimension, paired up with other close by gravitons connected by a line of energy between each two single gravitons, and started to rotate around each other, changing form to become the building blocks of our universe matter.

Ours is a four dimensional universe, containing three dimensions of space and one dimension of motion. The motion of matter gives it the energy to counter the ever present force of gravity, just like a bicycle without motion will fall. The motion of matter also gives it the energy to occupy different points in our universe space, and in this way, time ticks.

All the matter in our universe is involved in many forms of motion. It is always spinning around itself. It circles around heavier bodies of matter and also shoots outward, away from the point of the Big Bang. As a result of the tug-of-war between the forces of gravity and energy, all these motions of matter (including photons) move in a circular motion and never in a straight line. This gives our matter universe its round ball shape.

Matter in motion has energy. The faster its motion, the more energy it has, and likewise, the greater its mass, the more energy it has. The energy only materializes when the matter accelerates, decelerates or changes direction. Then, the energy transfers from one body of matter to another.

For example, a jet airplane takes off by burning its fuel in a chemical reaction with oxygen in the air, as the liquid fuel explodes and turns into gas. In other words, the molecules of which the liquid fuel is made need much more space than they require in their liquid state, and the movement of these molecules creates energy and pressure in all directions. The way the jet engines are configured only allows this exploded mixture of fuel to go backward, pushing the engine and connected airplane forward.

Most of the energy is needed to accelerate the airplane to its cruising speed of about five hundred miles per hour. Once it reaches that speed it still needs energy to counter Earth's gravitational pull, which it does by pushing air molecules downward with its wings. The resistance of these air molecules in the path of the airplane's wings to accelerate downward lifts the airplane upward and keeps it at the desired altitude.

The airplane also needs energy to counter the friction of air, which is the resistance of air molecules in its path. The airplane needs to change the speed and direction of those air molecules away from its body as it moves through the air. All this energy generated from the airplane fuel is therefore not lost, but is transferred to different bodies of matter, all the way down to atoms, in the form of heat.

Heat is another form of energy. It causes atoms to rotate and spin faster, the electrons that circle their nuclei to go into higher orbit, and the matter that contains these atoms to increase in size. The more energetic atoms produce more photons, which are themselves energy units. By releasing photons, atoms cool down. In turn, if these photons hit other atoms, they add their matter-energy to those atoms in the form of heat, accelerating the atom movement and adding mass. Thus, heat can be created by adding photons, or by the friction and collision of atoms with other atoms or subatomic particles. In other words, energy always involves matter in motion, which means that matter is itself a concentrated source of energy.

The liquid fuel we put in a car has enough chemical energy to accelerate the car to a certain speed for a certain distance. The chemical energy in the liquid fuel transforms into motion-energy of the car mass, and because of this increased motion the car has more energy.

To decelerate the car we apply the brakes. The friction between the rotor and the brake pads creates heat, causing a large number of the atoms from which the rotor and the brake pads are made to disconnect from their surfaces and accelerate outward. In this way the brake pad and the rotor change structure, becoming smaller.

Such a change in any structure involves energy (some structures will need more energy to change shape than others), and that energy always has to come from somewhere else. Chemical energy is the ability to release some or all of the energy it took to create that chemical through chemical reaction. The chemical energy that fuel contains does not disappear; it just changes form into motion and eventually into heat. The matter of the liquid fuel also doesn't disappear; it just changes form to become gas and ash.

The faster a car moves, the more energy it has in the square of its speed. In other words, a car traveling sixty miles per hour will have four times the energy of the same car traveling thirty miles per hour; not twice, so buckle up! An asteroid weighing one million tons and traveling at a speed of forty thousand miles per hour will have not twice, but four times the energy of an asteroid that weighs half a million tons and travels at the same speed. This energy materializes when the asteroid hits another matter object, like a planet, and the impact energy transforms into heat and the acceleration and deceleration of matter particles that were involved in the impact.

So, the heavier an object and/or the faster it travels, the more energy it will have. This means that matter objects traveling at infinite speed will have infinite energy, but this is not possible because impact with infinite energy means total devastation. To put order in our universe God set a speed limit, and that is the speed of light, which is about 186,000 miles per second (in a vacuum).

The maximum limit of speed also sets a limit on the maximum energy of matter. The maximum energy that matter can contain is its mass multiplied by the square of the speed of light, which is what Einstein's famous equation $E=MC^2$ is all about. "E" is the energy that equals mass, "M", times the square of the speed of light, "C", which is 186,000 miles per second. From this equation we understand also that mass equals energy by switching the equation to $M=E/C^2$. Matter is energy and it always involves motion.

Using an atomic bomb we can transform five pounds of uranium to one pound of pure energy in the form of photons. One pound of uranium traveling at the speed of light has as much energy as one thousand tons of TNT.

A sun fuses hydrogen atoms in its core under the tremendous pressure of its own gravitational force. Five atoms of hydrogen produce one atom of helium, but the total weight of one helium atom is less than the total weight of five hydrogen atoms. This missing weight transforms into photons, pure energy that shoots out from the sun's core in all directions. The photons created in the sun's core collide with all the atoms in their path, adding their mass energy to those atoms and causing heat, while in turn, these heated atoms produce photons to cool off. This chain reaction of photons continues outward from the sun's core, in a trip that can take many months, until the photons eventually leave the sun's surface. That's what we see when we look at the sun shining with all its glory.

Our Sun produces energy by fusing hydrogen atoms in its core and this energy offsets the gravitational force of the Sun's mass. By transforming matter into energy (photons), the Sun loses a few thousand tons of matter per second, but it has enough hydrogen atoms in its core to last another four billion years as a shining star.

So, energy manifests itself physically as matter in motion. The faster the motion of the matter the more energy it has. The maximum speed of motion is the speed of light. Only photons can travel at the speed of light. Therefore, photons are the most energetic particles. Photons are the pure energy which we refer to when we say that all the matter in our universe was created in the Big Bang from pure energy $M=E/C^2$.

7. Time-Space

Our universe was created in the Big Bang, the explosion/splitting of matter of a single dimension into a huge number of gravitons. This number, 10^{2000}, of single gravitons expanded to occupy an emptiness of about thirty billion light-years in all directions away from the point of the collision. This happened in no time, defining our universe space and borders by way of the presence of these single gravitons. With their motion, numbers and the distance between them, single gravitons impose speed limit and gravitational force on all matter structures, including photons. The single gravitons are in a different dimension and are subject to different laws of nature.

Some gravitons paired up at the point of the Big Bang where single gravitons were denser to create the building blocks of matter, a two-dimensional matter (like two dots connected by one line of energy). By grouping up with other pairs of gravitons in different numbers and shapes, they created all the diverse subatomic particles, and matter became three-dimensional.

The concentration of all those pairs of gravitons to create lumps of matter in a small space created imbalance between positive and negative energies in the fabric of our universe, leading to the Gravitational Force Phenomenon, wherein the motion of matter gives matter the energy to offset the motion of gravitational force.

The interplay between gravitational forces and energies always involves the motion of our universe matter. This motion is the Fourth Dimension of our universe. The fourth dimension, which causes all matter structures in our universe space to change position relative to all other matter structures and to the fabric of our universe itself, is the essence of **time**.

The current position of matter structures gives us the Present Time, its previous position is the Past and its future position is the Future. This means that when the relative position of matter structure changes, time ticks onward toward the future. If the relative position of matter structure stops changing, time stops. And when the relative position of matter structure goes back to what it used to be, time goes back also.

Therefore, time does not flow by itself; it's always connected to the relative position of matter structures. Time is only the reflection of the relative position of matter structures.

Matter structures are always in motion. They spin around themselves to counter their own gravitational force, they circle around objects of heavier mass to counter a stronger gravitational force, and they shoot away from the point of the Big Bang, and by doing so change their relative positions.

Matter is made from a huge number of structural units, smaller ones within bigger ones, and they are all in motion all the way down to their building blocks (the pairs of gravitons that are also spinning around themselves). Because of this motion, these structures of matter change their position relative to one another and also relative to the fabric of our universe, never occupying the same relative position more than once. This is the reason why time as a whole only flows forward.

We can create the illusion of time going backwards in a small number of matter structures by applying energy to move them backward to a previous position. However, this is only an illusion. As the smaller structures within the structure change position, the matter structures in the rest of the universe also change position, and are no longer in the same position relative to the fabric of our universe. Furthermore, there isn't enough energy in our universe to reposition all the matter structures that exist to a previous state. This is because we are always irretrievably losing matter-energy to black holes. So, time as a whole only goes forward toward the future, while the past is sealed and does not exist anymore.

It is not possible to go backward in time, except in your imagination. Where are all the smart future generations of humans with time machines that can go backward in time? I do not see them.

There is a reason why the matter structures in our universe are positioned and move in a certain manner and order. It is because of our nature's laws: the forces of energy and gravity and their byproduct magnetism. The same forces that shaped our past and shape our present will also shape the future. Therefore, the future is sealed, although it does not exist yet.

It is impossible to predict the future or even the past in great detail because the necessary information involves a huge number of interactions of matter structures moving in a very complex way, all according to the laws of our universe. Time always ticks forward because matter structures are always in motion, changing position relative to other matter structures and relative to our universe fabric, connecting and disconnecting with other matter structures and moving throughout our universe space.

Our universe space (fabric) is made up from such a huge number of single gravitons that the space between them is about one with forty zeros behind it of an inch, much smaller than the nucleus of one hydrogen atom. These single gravitons are repelled from each other by a line of negative energy that holds the single gravitons in their space. The distance between single gravitons is very small and it represents the smallest unit of time, which is the duration it takes matter to jump from one graviton to the next one. This is a very small space that only one pair of gravitons (one building block of matter) can jump into, moving along the negative energy line between the single gravitons.

The space between the single gravitons acts like a black hole. When there is an empty space there is no gravity and matter explodes. Matter is energy, and if there is no gravity to contain it, it will explode to become a different dimension that travels at a speed above the speed of light until it hits the next graviton. At that point it stops, reconstructs and becomes our universe matter again. Paradoxically, the space between the single gravitons takes no time to cross. Matter loses its identity as the matter of our universe as it travels along the negative energy lines between the single gravitons that are part of the fabric of our universe. Matter travels at a speed above the speed of light, which for us is right away (no time). Time restarts when this energy once again becomes matter as it hits the next graviton. And this is the smallest unit of time. This is our present time, when time stops because all the matter structures in our universe stop in one position relative to each other and also relative to the fabric of our universe. As you see, the real present of time is very small, one with eighty zeros behind it of a second (10^{80}). It is the duration that a photon appears on a single graviton before it jumps to the next graviton, a very small unit of time that is very hard to comprehend and impossible to measure.

So, to measure time we come down to Earth, literally. We divide into twenty-four units of time the duration it takes Earth, our home planet, to make one complete spin around its axis relative to the position of our Sun, and we call these units of time "hours". Each hour is split into sixty parts and we call them "minutes". Each minute is split into sixty parts again and we call those units of time "seconds". We can divide each second into fractions, all the way down to one with eighty zeros behind it of a second, which is the smallest unit of time. This unit of time is when time stops because all the matter units in our universe stop in one position relative to each other and also relative to the fabric of our universe (the web of a huge number of single gravitons that occupies our universe space).

It is the motion of matter throughout our universe space that gives us time. But the motion (speed and direction) of matter varies, depending on its energy and the energy that surrounds it, including gravity. This variability means that we cannot measure time very accurately. For example, let's go back to Earth again: one complete spin of Earth occurs every twenty-four hours, but every day Earth is subject to forces that could accelerate or decelerate its spin by a fraction of a second each day. These forces could be weather patterns, volcanic activity, or bombardment by space objects such as solar wind and asteroids. Earth's spin is also influenced by the gravity of the Moon, Sun and other planets in our Solar System as they change their positions relative to Earth.

We build clocks and watches to correspond to the speed of Earth's motion, but their movement is also influenced by magnetic fields, batteries, springs, wear and tear, temperature and atmospheric pressure. Even atomic clocks are not exempt. The spin of atoms in the heart of an atomic clock is variable depending on the amount of photons they receive or discharge, on gravity and on the overall speed of the atomic clock unit itself through space. As with everything else, it has a finite lifespan and will wear out eventually.

We know that time does not move alone and is always connected to the position of matter structures in our universe space, as a reflection or a shadow of the position. This is true for the position of all matter objects, from the very small, like the pairs of gravitons that are the building blocks of all matter, to the very large objects like planets, stars and galaxies. All those matter structures are always in motion changing position, and time ticks on for them, which means that time and matter always go hand in hand.

When matter is born its time is also born, and when the matter dies its time also stops. Let me explain this in more depth. Matter is made from a huge number of structures all the way down to the pairs of gravitons, and each one of these matter structures has its own time ticking for it, as long as it is changing its relative position. Some of those matter structures group together to form a unique body of matter with its own energy and motion, and time is born to this new structure of matter. Once this body of matter ceases to exist its time also stops, because this unique structure of matter does not exist anymore. So time itself is born and dies. It has a beginning and an end because it is always connected to matter structures, and matter structures always have a beginning and an end, or to put it more correctly, they change form to become different structures.

Time has a beginning and an end just like matter structures in the circle of life. For example, asteroids (those big chunks of rocks made mostly out of heavy metal) are born from the collisions of brown and black dwarfs or planets, and their time is born with them because they don't exist before the collision that creates them. Asteroids can survive for billions of years because they are a unique structure of matter. They have a presence, they are in motion, they change position and time ticks on for them. But once they hit a planet or a sun they disappear, changing form to become an integrated part of that sun or planet, and time stops for them. In this example, time moved full circle; it was born with the asteroid in the first collision and died with the asteroid in the last collision.

A baby is born as a result of a huge number of interactions of small structures of matter that change form to form a very complex and larger structure of matter: a human baby. When the baby is born its time is born with him. He didn't exist before, and therefore he didn't have any time. Time starts to tick for him when he is born and he goes through the cycle of life changing forms all the time, which we perceive as getting old. This baby becomes an old man when he reaches the ripe age of one hundred and thirty, because he read this book of knowledge and took care of himself. Finally, after circling the Sun one hundred and thirty times since his birth, his body can no longer sustain life because of the continuous changes in his body. Time stops for him as a living man and restarts again as a decomposed body that goes through another cycle of matter structures, changing forms to become a different matter structure.

Changing structures is another hallmark of time because time ticks forward toward the future at the speed of one with eighty zeros behind it times a second, which is the smallest unit of time. This is our present time, when time stops and restarts, and that means that matter structures move at the speed of one with eighty zeros behind it each second. They change their relative positions, and therefore, the whole structure changes.

So how old are you? Fifteen billion years? That is when the matter in our universe was created from a different dimension and your body was made from this new matter. Are you five billion years old? That is when our Solar System came into existence and your body became a part of it. Are you twenty-eight years old? Did the Earth circle the Sun twenty-eight times since your body was put together in your mother's womb? Or are you one with eighty zeros behind it of a second old, which was the last time that your body structures changed their positions relative to the rest of our universe, meaning you are not the same person that you were before?

The answers to all those questions is: yes indeed. Time is only a reflection, an illusion of the brain, and it is not real. Yet time is everything. It is not only what you own, but is how long you own it, be it your house, your money, or your body. You cannot hold time and it does not flow like a river. You cannot go backward or forward in time. The past is gone and it is not there anymore, because the relative positions of the matter structures changed. The future is not anywhere yet, so there is no place to go.

We perceive time as the relative position of matter structures to other matter structures, and in order for them to change relative position they have to have motion, which is the fourth dimension. The slower the motion of matter structures relative to their surroundings, the slower their relative time ticks, because it takes them longer to occupy different points in space, and for that we need empty space between points. Once the motion of matter structures stops from within, their own relative positions stay the same and time stops for them, meaning their time does not flow and they stay in the present state of time. For example, if you sit on a chair for one hour, the relative position of your butt to the chair is the same for one hour. Time will stop for you relative to the chair for one hour because the relative position between you and the chair is unchanged; you become one structure for one hour. Therefore, the present time can last as long as the relative position is unchanged.

The huge number of objects in our universe are always in motion. Motion gives them energy to counter the force of gravity and to occupy different points in our universe space. Those points are the huge number of single gravitons that define our universe space and boundaries. The empty space between the single gravitons is the empty space we need for time to tick.

Matter is energy, which means that any matter structure leaving our universe space will explode when it passes our universe boundaries into emptiness, where there are no gravitons to contain its energy by gravity force. The exploded matter will become a different dimension and time will stop for it. The first reason time will stop is that it is not matter as we know it. The second reason is that there are no points in the empty space to which it can relate its speed, which means that its speed could be infinitely fast or zero, either of which will cause time as we know it to stop.

Our universe time started when matter was born, because we always connect time to the relative positions of matter. But our universe-matter was born from a different dimension, and therefore, existed in a different dimension of time. Furthermore, if we compare our universe time to the infinite universe time, which is infinite in size as well as timeless because it has no beginning and therefore no end, we get two answers. First, we come out with an infinitely small number (infinitely small number means smaller than any number you can imagine), which means our universe time is infinitely small, or in other words, a blink of the eye for the infinite universe. The second answer is:

zero. This means our universe time is an illusion, that there is nothing to hold on to. So time for us is only the relative position of matter structures to other matter structures in our universe all the way down to the smallest unit of matter, the pair of gravitons that spin around themselves. The shorter the distance between these two gravitons, the slower their spin, and the slower time ticks for them because it takes more time for them to change position (to spin around themselves relative to other pairs). If there is no distance between these two gravitons, then there is no spin, and time stops for them as it does for a photon.

Why is time an illusion? Only God knows. (For the answer see the chapter "God and The Illusion of Time.") I will speak more about "time" in the next chapter, "The Constant of the Speed of Light."

8. The Constant of the Speed of Light

God created the speed limit in our universe to impose some kind of order within it. Because energy comes with the motion of matter and with infinite speed, there will be infinite energy and chaos that can lead to infinite possibilities. So to put an order in our universe, God set up a speed limit, which is about 186,000 miles per second.

God created this speed limit in a beautifully simple way. He does not have time for complex things. His time is wasted by all those crazy people knocking on Heaven's door and demanding first class service.

God created our universe by splitting a one-dimensional point into a huge number of gravitons (which are by themselves one-dimensional points) in an empty space so that they occupied this space as a web made of a huge number of points thirty billion light-years in all directions away from the point of the Big Bang. This defined our universe space, borders and properties.

Therein lies the simple beauty. This web of single gravitons throughout our universe imposed gravity force, space and speed limit, and therefore time itself on all matter structures, including photons. This is simply due to the number of single gravitons, their motion and the size of the empty spaces between them. The empty spaces in between the single gravitons are occupied by negative (repulsive) energy lines that repel the single gravitons away from each other, since the gravitons are positive energy units that are attracted to each other.

Each building block of matter is made from two gravitons with a space between them. These two gravitons spin around themselves connected by a line of energy. This is two-dimensional matter that becomes three-dimensional when the pair connects to other pairs. The fourth dimension is the motion of all these matter structures which gives them energy to counter gravity force and also to occupy different points in our universe space. By doing so, they change their relative positions and time ticks.

Photons are also made from pairs of gravitons, but unlike the building blocks of matter, there is no space between the two gravitons that form a single photon. Therefore, photons are one-dimensional, and it is their motion throughout our universe space that gives them their second dimension. These pairs of gravitons that make photons are glued together and become one. They do not spin around themselves and therefore their relative position is unchanged. This means that time stops for them. They are always in the present state of time and they never change form from the moment of birth until the end, which can take anywhere between a split second to many billions of years. They change form only when they hit a black hole or a matter structure and add their mass-energy to it.

The motion of all matter throughout our universe space, including photons, is itself an illusion. This is because in the empty spaces along the negative energy lines between the single gravitons that define our universe space all matter structures and photons disappear, exploding to become a different dimension, at which point time stops ticking for them. Once the energies of this different dimension hit the next graviton in their path, they become our universe matter again and cause time to restart as the present time.

We get the same illusion of motion when we watch moving pictures on the television screen. We really do not see motion. We see only different stationary dots (pixels) on the television screen that light up and turn off in sequence to give us the illusion that the picture is in motion. The web of gravitons that occupies our universe space works on the same principle. The building blocks of matter and the photons disappear in the empty spaces between the single gravitons and reappear when they hit the next graviton.

The duration for which the matter (including the photons) appears on the single gravitons before they jump to the next graviton is the present time of our universe. This is when all the matter structures in the universe are in one position relative to each other and also relative to the fabric of our universe, the web of single gravitons. The duration for which matter structures appear on single gravitons depends on their energy, or in other words, on their speed of motion. The faster the motion of a matter structure, the more energy it has.

This higher energy of motion in matter manifests itself physically through the change in the structure of the building blocks of that matter. When the distance between the two gravitons that are the building blocks of all matter becomes shorter, it shortens the duration of time for which they will appear on the next single graviton in their path. The shorter the distance between the two gravitons that form a single block of matter, the faster they jump from one graviton to the next. This also means that they're going through greater numbers of single gravitons in their path in a given amount of time. These single gravitons are gravity force units and therefore matter will become heavier the more single gravitons they pass per given amount of time.

According to Einstein's equation $M = M_0 / \sqrt{(1-V^2/C^2)}$ matter structures become heavier the faster they move up to the speed of light when their mass becomes infinitely heavy and therefore we cannot accelerate their speed to the speed of light or beyond because we will need infinite energy. This equation shows the genius of Einstein and the limitation imposed on our universe in a mathematical language. But this equation deals with motion (speed) and as it's shown earlier in this chapter motion is an illusion. Matter structures, including photons appear, disappear and reappear again on the next single graviton in their path. So, we have to tinker with this equation to match reality, but the end result will be the same: matter structures will become heavier the faster they move/jump through our universe's space--$M=M_0/ \sqrt{(1-V^2/10^{80\ 2})}$. "M" is the mass which depends on the number of building blocks of matter (the pairs of gravitons) of which this matter structure is made equal "M_0" the same mass at rest divided by the square roots of inside the parenthesis (One minus "V^2" the square number of single gravitons this matter jumps per second relative to the other system divided by the square of "$10^{80\ 2}$" the number of single gravitons the photons jump per second, which is constant).

So, the more single gravitons matter structures go through in a given time period, the heavier they become and the shorter the space between the matter structures becomes, all the way down to the building blocks of matter (the pairs of gravitons). When the distance between them gets shorter their spin becomes slower, they change their relative positions more slowly, and therefore, time ticks more slowly for them relative to normal matter structures that have a lower velocity.

We see exactly the same effect when matter structures are under gravity force. Under gravity force single gravitons stream into the weak gaps (the weakening and shortening of the repulsive energy lines in the gaps between single gravitons) that all matter structures create by way of their presence in the fabric of our universe. The more concentrated a matter structure, the more weak and short the gaps it creates in the fabric of our universe, and the faster the single gravitons nearby move towards the short gaps, carrying any matter structure along with them for the duration the matter structure appears on them, which leads to a stronger gravity force that squeezes the matter to its center. As a result, the matter structures get smaller and heavier while their relative time slows down.

So, the only difference between matter structures in motion and matter structures under gravity force is that matter structures in motion hit the single gravitons in their path, while single gravitons fall and hit matter structures under gravity force. Both gravity force and high velocity have the same exact effect: matter structures get smaller and heavier while their relative time gets slower.

For example, one cubic meter of water weighs one ton on Earth. The same cubic meter of water weighs one-sixth of a ton on the Moon, which is only about one hundred and sixty-six kilograms. This is due to the fact that the Moon has only one-sixth of Earth's gravitational force. The same cubic meter of water weighs thirty tons on the Sun's surface because the Sun has thirty times the gravitational force of Earth and will therefore need nine hundred times the energy (the square of thirty) to lift this cubic meter of water relative to the energy needed on Earth. The stronger gravity force on the Sun is due to the faster rate at which single gravitons are falling toward it.

The same thing happens at a higher velocity. The faster matter objects move through single gravitons, the heavier they become, just as if they were under gravity force, and we will need more energy to accelerate them in the square of their weight. If we take a one hundred foot poll from Earth and stick it on the Sun's surface, the pole will become about one inch shorter. This is because the Sun has a stronger gravity force than Earth that compresses matter structures as single gravitons fall in toward the Sun at a faster rate than they fall toward Earth. The same thing happens at a higher velocity through our universe space. The faster matter objects move through space, the more gravitons will squeeze them and make them shorter, just as though they were under a stronger gravity force.

On the other hand, all matter objects will stretch (get longer) in a free fall toward any object that creates a gravitational field, depending on the force of gravity, as they get carried by the fabric of our universe, since our universe space moves faster in front of matter structures than it does to the rear of them.

So, upon encountering higher speeds or stronger gravity all matter structures become smaller and heavier and their relative time slows down. At the speed of light, which is about 186,000 miles per second, all matter structures have to go through one with eighty zeros behind it of single gravitons in their path per second. At that speed the single gravitons will squeeze the building blocks of any matter structure into single, or one-dimensional, points. In other words, the building blocks of matter (the pairs of gravitons) will lose all the space between them to become photons. Time will stop for them because the pairs of gravitons of which each photon is made will now have no space between them and therefore no spin, meaning their relative positions will stay the same.

We get the same effect on the surface of a black hole, which only photons can reach because the same numbers of single gravitons the photons pass at the speed of light are falling in toward them at the surface of the black hole. This means that the gravitons that carry gravity force and are part of the fabric of our universe also have speed limit, which is the same as the speed of light, and they only reach that speed when they hit the surface of a black hole. Therefore, the ring around a black hole is a ring of single gravitons that are part of the fabric of our universe and a ring of photons that cannot escape because of the single gravitons hitting them at the speed of light with the same number of gravitons that each photon will hit per second at the speed of light. So, for us the photon will seem as if it is motionless, but for the photon it will seem as if it is traveling at the speed of light, since it gets hit or passes the same number of gravitons.

Light is made from a huge number of photons and their speed is constant at a speed of about 186,000 miles per second. But we measure speed by how much distance matter objects move in a given amount of time. For example, if we drive a car at the speed of sixty miles per hour, we will travel one mile every minute. So we put a marker on the highway, a poll for every mile. If we drive for one hour we pass sixty poles, from which we can deduce that we drove sixty miles and that our speed was sixty miles per hour.

Now, what if somebody without our knowledge increased the distance between the poles by ten percent? If we drove on the same highway for one hour and we passed sixty poles, we would think that we drove sixty miles and that our speed was sixty miles per hour, whereas in reality we would have travelled sixty-six miles at a speed of sixty-six miles per hour. Although the distance between the poles increased by ten percent, we measured our distance by the number of poles we passed in one hour and not the real distance.

We have the same problem with our universe space in that the distance is not absolute, but is actually the number of points in space, and those points are gravitons. These single gravitons can have different distances between them and are also in motion all the time, so distance in space is not absolute. We have the same problem with the speed of light, which is constant, because although a photon goes through the same number of single gravitons per second, the distance between the single gravitons fluctuates depending on gravity and the space-time of our universe itself.

For example, shortly after the Big Bang and the creation of photons, the space between single gravitons near the point of the Big Bang was shorter than today's space by a few percent. Therefore, photons going through the same number of single gravitons in their path per second appear as if their speed decreases by a few percent because the total distance those photons are going through per second decreases.

Furthermore, the photon, just as any matter structure, will increase or decrease its total speed as it is carried on the surface of a single graviton for the duration it appears on that graviton. This depends on the directional motion of the graviton itself and on the directional motion of the photon relative to any object surface, but not relative to the number of gravitons they pass. In other words, the photons will appear, disappear and reappear on the same number of gravitons in their path because our universe space is really the number of single gravitons and those gravitons are always in motion.

For example, photons leaving our Sun's surface will hit the surface of Earth in about eight minutes. Therefore, those photons cover a distance of about ninety million miles (eight minutes travel at the speed of light). However, photons leaving Earth's surface will hit the surface of the Sun in about three seconds less, which means that those photons cover a distance of only about eighty-nine million five hundred thousand miles. This is because the Sun creates a much stronger gravitational field than Earth and the movement of gravitons towards the Sun is from a longer distance and at a faster rate than towards the Earth.

This makes the space shorter going towards the Sun and longer leaving it.

When I was a kid growing up in Israel a question came up: "Does Israel have more upward hills than downward hills?" I answered "upward hills," since it is harder to go uphill. You need more energy and/or time to go uphill. Therefore, upward hills seem longer than downward hills. This was the wrong answer for that time, because there are exactly the same number of hills that go up as there are hills that go down, because you can go up and down the same hill. But now, armed with the knowledge of the workings of our universe, I know that I was right and that there are more upward hills. The actual space is longer uphill because you move against gravity force, just as you would if you were walking against an escalator.

We can test this new theory concerning the fluctuation of distance or the variable speed of light. Both depend on the directional motion of single gravitons in that the distance will seem longer going against the directional motion of single gravitons, and the other way around. The speed of any object (including photons) will decrease when moving against the directional motion of single gravitons, and it will increase when moving with the directional motion of single gravitons. We can use a spacecraft packed with sensitive instruments, including an atomic clock. The spacecraft has to be in high orbit and synchronized with Earth's motion. We need to put the same instruments below the spacecraft, on Earth's surface, including an atomic clock and taking into account that the atomic clock on Earth's surface will tick more slowly than the one in the spacecraft above. Now we shoot photons at exactly the same time from both places towards each other. What we will find is that the instruments on Earth will receive the photons coming from the spacecraft about a fraction of one second faster than the spacecraft will.

Space (distance) can shrink down to zero or expand to the speed of light, depending on the speed and direction of the single gravitons themselves. This is the reason a photon could never leave the surface of a black hole: the same number of gravitons the photon would pass at the speed of light are falling toward it at the surface of a black hole. This also means that a photon going straight toward a black hole reaches twice the speed of light when it hits the surface of a black hole. For example, let's take an airplane that reaches the speed of two hundred miles per hour. If this airplane flies against a two hundred mile per hour wind, its speed will be zero relative to Earth. The same airplane flying in the same direction as the wind will have a speed of four hundred miles per hour relative to Earth. In both cases the airplane's speed relative to the air is two hundred miles per hour. The air, and not the Earth, is the airplane's space.

So, the maximum speed of a photon is twice the speed of light as it falls toward a black hole relative to the surface of the black hole, but relative to its own space its speed is always the speed of light. This is just like any object falling toward any planet. The speed of the object will increase relative to the planet's surface, but the falling object won't feel any acceleration because its space is moving at the same rate of speed as the falling object. The object will feel only the weightlessness, but no acceleration.

The speed of light (photons) is therefore constant in that it always passes through the exact same number of single gravitons in its path per second, but the distance light travels, in terms of distance as we understand it, varies. Therefore, space/distance is not absolute. It is an illusion because space is only a certain number of single gravitons. These single gravitons are always in motion and the space between them is not constant; it fluctuates.

The empty space between single gravitons is a "no time zone" because the building blocks of matter and photons jump from one graviton to the next in no time to reappear on the next graviton, and time restarts as the present time. The duration they appear on a graviton depends on their energy of motion, or more correctly, on the distance between the two gravitons of which the building blocks of matter are made. The longer the distance between the two gravitons in the pair, the longer the duration of time they appear on a single graviton. The shorter the distance, the shorter the time, and they jump to the next graviton faster. A photon has the most energy because it is made from a pair of gravitons with no distance between them, and will therefore have the shortest appearing duration possible when it hits the next single graviton in its path. Its velocity is about 186,000 miles per second and its speed does not change as it goes through the same number of single gravitons in its path per second. There is no friction and the single graviton will put out the exact energy it receives. Therefore, the speed will be unchanged.

The same is true for matter structures. Their initial speed will stay the same as they go through the fabric of our universe with no friction. To increase the speed of matter structures we need energy. The faster the matter structures move, the heavier they become and the more energy we need to accelerate their speed, while their size shrinks, their relative motion slows down, and their time ticks more slowly. All those changes are profound when they reach speeds close to the speed of light, or when matter structures get closer to the surface of a black hole as they move through a huge number of single gravitons.

The same amount of energy needed to accelerate the speed of a matter structure is likewise needed to decelerate the speed of a matter structure. The same amount of energy that is needed to shorten the distance between two gravitons (of which the building blocks of matter are made) is also needed to extend that distance.

There is a direct correlation between the distance of two gravitons from each other to their speed, or their motion, throughout our universe space. The faster the building blocks of matter move through our universe space, the more single gravitons they hit in their path, the shorter the distance between the pair of gravitons becomes, the heaver they become, the more slowly they spin, the more slowly time ticks, and the lower their temperature becomes.

We get the same exact effect when the building blocks of matter are under a strong gravity force. The more single gravitons fall on a matter structure, the shorter the distance between the pairs of gravitons of that matter structure becomes, the heavier they become, the more slowly they spin, the slower their relative time, and the lower their temperature becomes. All this means is that just by knowing the distance between two gravitons we can figure out the temperature of a matter structure, its relative time and its energy, which are in direct correlation to its speed, that is, its motion through the web of single gravitons. We can figure out the amount of gravitational force to which this matter is subject, or in other words, how many single gravitons are in its path per second or falling toward it per second, which will be the same at certain speeds or at certain levels of gravity force.

For example, at the speed of light or at the surface of a black hole the number of single gravitons hitting the photons will be the same. This also means that gravity force is the mirror image of energy and that gravity force and energy of motion are basically the same thing. All energy comes from matter structures in motion through our universe space. The faster they move, the more energy they have, and the more single gravitons per second the matter structures will hit. Gravity force is how many gravitons per second hit a matter structure. The more gravitons hit a matter structure per second, the greater the force of gravity becomes.

It is like you are riding a quiet train at thirty-six kilometers per hour and alongside the train there are trees ten meters apart, so every second you pass a tree as the train goes forward. Now close your eyes and imagine that the train is at a standstill, and open your eyes again. You will see that the trees are passing you at a rate of one tree per second. Thus, energy and gravity are the same. They are simply opposite directions. It only depends on how you look at it.

Spin is another form of motion. All matter structures spin, this gives them the energy to offset the inward force of gravity. The heavier they are, the more concentrated they are, the stronger their gravity force and the faster they spin, while magnetism, which is the byproduct of spin, also becomes stronger. Clearly, there is a direct correlation between magnetism, mass, energy and the motion/direction of gravity.

We can measure gravitational force by the speed of light. We know that the speed of light is constant and that the photon speed is always the same relative to its space, which is made of a huge number of gravitons constantly in motion. However, the speed of the fabric of our universe is not constant. It ranges from close to zero to the speed of light, and it depends on the distance of the single gravitons (which are part of the fabric of our universe) from any matter structure or black hole surface and on the total mass and size of those matter structures or black holes.

At the surface of a black hole those single gravitons will fall in at the speed of light and will carry photons with them, which always move at the speed of light. This means that the photon speed will be twice the speed of light when it hits the black hole surface, relative to the black hole surface. This speed is the photon's speed going straight forward towards the black hole surface combined with the speed of single gravitons moving at the speed of light when they hit the black hole surface carrying the photons along with them for the duration the photons appear on them. So, the photon speed relative to the black hole surface will be twice the speed of light, but the photon speed relative to its space, which is made from single gravitons, will be the speed of light.

On the other hand, photons moving away from the surface of a black hole have a speed of zero relative to the black hole surface, but the speed of light relative to its space means that space itself is in motion falling toward the photons at the speed of light. This means the surface of a black hole has a gravity force equal to one hundred percent of the speed of light, and the photon cannot escape this gravity force.

The surface of a neutron star has a gravity force equal to ninety percent of the speed of light. This means that the speed of photons hitting the neutron star's surface is one hundred and ninety percent of the speed of light and ten percent escaping it. Our Sun's surface has a gravity force equal only to a few percent of the speed of light, and Earth has only a fraction of one percent.

To make things more complex, the atomic nucleus has the same gravity force as a neutron star on a much smaller scale.

All this means is that the speed of light is constant relative to the fabric of our universe, but that we have to add or deduct the motion of the fabric of our universe itself to the constant of the speed of light depending on its direction because of the ever present force of gravity.

So, the distance between two gravitons that make our building blocks of matter can tell us the speed of the matter or to what amount of gravity force the pair of gravitons are subject, as well as the relative time of the matter structure, since time is only the reflection of the position of a matter structure relative to other matter structures. As the distance between two gravitons becomes shorter, their spin becomes slower, they change their relative position more slowly, and this means that time ticks more slowly for them relative to other building blocks of matter that are spinning more quickly because the distance between their graviton pairs is greater.

The same is true for temperature: the shorter the distance between the two gravitons in the pair (the building blocks of matter), the lower the temperature. We perceive heat as the amount of heat matter structures produce in the form of photons: the more photons a matter structure produces, the hotter it is. All matter structures produce heat (photons). It is a byproduct of their building blocks, the spins of pairs of gravitons and their collisions with other pairs of gravitons to create pairs of gravitons with no distance between them, which are photons, and this means they also lose mass-energy and become colder. So, the longer the distance between the two gravitons that form a single building block of matter, the faster they spin and the more collisions there are, which in turn means they will produce more photons and become hotter. The shorter the distance between two gravitons, the slower their spin and the less collisions they produce, meaning there will be less photons and as a result lesser heat and lower temperatures.

To measure how much less heat a matter object will emit in the form of photons under gravitational force or high velocity relative to the same matter object at rest, we can use Einstein's equation that describes time dilation (the slowing of time). Since this equation really describes the relative slowing of the internal motion of the matter object all the way down to the spinning motion of the building blocks of matter (the smallest unit of matter) and not time. Time is an illusion. Motion triggers the illusion of time ticking forward (the faster the motion the faster the illusion of time ticking forward, and conversely, the slower the motion the slower the illusion of time ticks forward).

Therefore, this equation will show us how much the spin motion of the building blocks of matter will slow down at higher velocity or at equal gravitational force. For example, our sun creates gravitational force equal to a few percent of the speed of light, and therefore, without this gravitational force that squeezes its matter the sun would be a few percent hotter. So, the smaller the distance between the two gravitons in the pair the slower their spin motion the less collisions of the building blocks of matter among themselves the less heat (photons) they produce and the colder they become down to the absolute zero and time stops—Matter Temp=$MT_0 / \sqrt{(1 - V^2/10^{80\,2})}$

At the speed of light matter structures will lose all their space and become either black holes or photons. Paradoxically, photons do not emit heat because they do not produce other photons. The two gravitons of which each photon is made do not have any distance between them and therefore do not spin, their relative position does not change, so they are always in the present state of time and their temperature is absolute zero.

So, in a simple and beautiful way God unified all the forces of nature (gravity, matter-energy, thermo-energy, magnetism and time itself) just by determining numbers of gravitons and the distance between them. This is true for the fabric of our universe, which is made from a huge number of single gravitons, and also true for the building blocks of matter, the pairs of gravitons whose intervening distance controls the way they interact with the fabric of our universe to create their speed and therefore their energy and mass.

The shorter the distance between two gravitons, the heavier they become and the more energy or gravity force is needed to squeeze them further down. The distance between two gravitons indicates how many single gravitons are in their path or how many single gravitons are falling toward them. Those single gravitons carry the gravity force and will make the building blocks of matter that pass through them heavier. Traveling at about ninety percent the speed of light the building blocks of matter will lose about half their size, they will be much heavier, their spin will be lessened by about fifty percent, and therefore, their time will tick about fifty percent more slowly and their temperature will only be a few degrees above absolute zero. Traveling at the speed of light the building blocks of matter will lose all of their space, their time will stop, their temperature will go down to absolute zero and they will become photons. It is impossible to accelerate a matter structure to beyond the speed of light unless it becomes a different dimension inside of a black hole.

Photons come out in a huge number of different wavelengths, that is, different lengths of space between the photons as the photons come out one after another. The shorter the wavelength, the more powerful it is. The shortest wavelength possible is our present time. That is the time in which all matter structures in our universe are in one position relative to each other and also relative to the fabric of our universe. That time is one with eighty zeros behind it of a second. It is the shortest wavelength possible and therefore has the most energy. As the photons come out as a wave, the distance between them is only one with eighty zeros behind it of a second, which is the duration any photon appears on a single graviton before it jumps to the next one.

At the time of the Big Bang, when single gravitons were denser, some of them collided to form pairs with no distance between them, and they became photons. The other way of making photons is through collision of the building blocks of matter among themselves. The collision of two building blocks of matter can produce one building block of matter (made of two gravitons connected by one line of energy) and one photon (made of two gravitons with no distance between them), or it can produce two photons.

So, it is wrong to assume that the distance between the two gravitons of which each building block of matter is made determines their speed through space. It is correct to assume that the distance between the two gravitons determines the number of the single gravitons in their path per second because space itself is an illusion. Space is only any number of single gravitons. Therefore, the smallest space-time is made of two single gravitons, regardless of the distance between these two gravitons. The space between single gravitons does not count as space-time for matter structures.

This space in between the single gravitons is the space of uncertainty where quantum mechanics takes effect. This space is where strange things can and do happen to small matter objects, mostly to photons. Once those matter objects disappear to become energies of a different dimension in between the single gravitons, they can materialize as our universe matter on the next single graviton in their path in no time. However, the next single graviton in their path could be a light-year away even though they may pass many single gravitons on the way without interaction. Or, these energies of a different dimension could split and appear on two or more different single gravitons at the same time, creating the same matter object at once.

As I mentioned earlier, the motion of the building blocks of matter and photons throughout our universe space is an illusion. They are not in motion, but in actuality disappear and reappear. So basically, the distance between each two gravitons that make up our building blocks of matter determines the speed at which they will appear, disappear and reappear again on the next single graviton in their path. The shorter the distance between the two gravitons, the faster they can appear and disappear per second. Therefore, photons (which have no distance between their pairs of gravitons) have the fastest rate of appearing and disappearing per second (about 10^{80} times per second), which is the constant speed of light (about 186,000 miles per second). We cannot determine this exact distance because space itself fluctuates depending on the space-time of our universe and on the forces of gravity (the direction of the single gravitons).

All this means is that the speed of light (the photon) is constant, and that it will always appear and disappear the same number of times per second on the surface of the single graviton in its path. For example, if a spaceship traveling at exactly half the speed of light shoots a missile with enough energy to reach ten percent of the speed of light relative to its starting point, the missile speed will be sixty percent of the speed of light. It starts at fifty percent of the speed of light, the initial speed of the spacecraft, and adds on ten percent. Now imagine that the same spaceship traveling at the same speed (fifty percent of the speed of light) shoots a light beam of photons at a certain wavelength. Knowing that photons always travel at one hundred percent of the speed of light and adding on the initial speed of fifty percent of the speed of light, can we conclude that those photons will travel at one hundred and fifty percent of the speed of light? The answer is: no.

Photons always travel at the speed of light regardless of their initial speed or the speed of their starting point. Photons always travel at the speed of light, which is about 186,000 miles per second, or more correctly, they appear and disappear the same number of times per second on the surface of the single gravitons in their path.

When we use energy to accelerate the motion of matter objects we do not really accelerate their motion because motion is really an illusion. We apply energy to bring the two gravitons of which the matter is made closer to each other, and the distance between those two gravitons determines their energy and the speed at which they will appear and disappear on the next single graviton.

The same thing happens with deceleration. We need energy to decelerate the motion of matter objects, and that energy goes toward extending the distance between two gravitons that make up the building blocks of matter. The distance between these two gravitons cannot be greater than the distance between two single gravitons that are part of the fabric of our universe. Once deceleration occurs through a huge amount of energy, the speed of the matter structure will be zero for a very short period of time relative to the fabric of our universe, because loose gravitons pair up almost immediately with other loose gravitons behind them. The pairing will result in two gravitons with no distance between them to become a photon traveling at the speed of light. So basically, we need the same energy to accelerate matter to the speed of light as we do to decelerate matter to a speed of zero relative to the fabric of our universe, and zero speed becomes the speed of light.

On the other hand, if we apply energy or gravity pressure to matter structures without accelerating their motion, the energy / gravity pressure expands the matter structures all the way down to their building blocks (the pairs of gravitons), causing the distance between them to increase and giving them a faster spin. This results in an increase of collision among those many building blocks (of which that matter structure is made) and thus in the production of more photons and an increase in heat. For example, as you walk on the ground the area beneath your footsteps heats up and produces photons that are visible in the infrared spectrum of light. The greater the pressure, the more heat it creates. This is the reason why it gets hotter as you dig deeper into the Earth (in miles), or why the neutron star shoots out an enormous amount of photons.

So, the closer two gravitons are in the pair that forms a single block of matter, the faster the rate of their appearing and disappearing will be on the fabric of our universe. The photon has the highest speed, and any additional energy to increase its speed won't work because the distance between the two gravitons that make a photon is already zero. The photon becomes one dimension, the closest any matter object can get to a single graviton. It becomes timeless and temperature-less.

So, time is an illusion, space is an illusion, and motion is an illusion. Life must be a sweet dream... Or is it?

9. Correcting Quantum Mechanics

Quantum mechanics deals with the very small matter object (the size of an atom or smaller) and therefore also with small spaces where it claims that the gravity force is very weak. And that is fundamentally wrong. Gravity is very strong where there is a concentrated (dense) matter structure occupying a small space, such as an atomic nucleus. In such conditions the gravity waves (gravitons) will move very fast (over ninety percent of the speed of light), carrying the matter with them for the duration the matter appears on them. This is what gravity is all about.

According to Isaac Newton's law of gravity, the closer matter objects are to each other, the stronger the force of gravity between them will be. He explained that the strength of their attraction will depend on "their masses divided by the square of their separation times a constant of proportionality." Only energy in the form of the spin motion can prevent these matter objects from colliding, due to the strong gravity attraction between them.

An atom's nucleus is made from many neutrons and protons (except for the hydrogen atom, which has one proton). Each neutron and proton is made from many subatomic particles and each particle is made from many building blocks of matter (excepting the photon, which has one building block of matter).

As you can see, the nucleus is a very dense matter structure resembling a neutron star's density. There is a very short distance between the many matter particles of which the nucleus is made.

All the matter particles that make an atom move and spin at high speeds (over ninety percent of the speed of light) because of strong gravity force. All this high speed motion in a very small space creates many collisions among the huge number of building blocks of matter comprising the nucleus (each proton or neutron is made from over a million building blocks of matter), and these collisions create photons. Collisions of photons with other photons or with small matter particles create new matter and anti-matter particles (anti-matter particles spin backward). A collision of anti-matter particles with matter particles creates photons. All this happens very quickly (close to the speed of light) and in a very tiny space, and therefore in a very small amount of time (less than a nanosecond). So, all these photons and new particles, including anti-matter particles that the atom creates, are real and not virtual. Most of them fall back to the atomic nucleus and change form, and this cycle continues, since the atomic nucleus is a powerhouse of energy where gravity is supreme.

Gravity is the force that prevents the atom's nucleus from tearing apart despite its fast spin. Gravity keeps it intact.

10. Black Holes and Beyond

Black holes are simply empty spaces where the fabric of our universe (our universe's space) is not present to transmit information and to impose control over matter structures through gravity force and speed limit.

The fabric of our universe is made from a huge number of single gravitons that are very dense. There are over one with eighty zeros behind it of single gravitons per one second of the speed of light, but there are still empty spaces between them. These empty spaces are occupied by lines of repulsive (negative) energy that keep the single gravitons at a certain distance from each other, since the gravitons are positive energy units that are attracted to each other.

Once a building block of matter appears on a single graviton, the negative energy line that repels the single gravitons from each other gets weaker and shorter in the area where the building block of matter appears, since the building blocks of matter are made from pairs of gravitons that are also positive energy units. Therefore, the other single gravitons around the single graviton that carries the most building blocks of matter move closer to it while carrying any matter structures with them for the duration the matter structures appear on them, which is the gravitational force moving and squeezing matter to its center or wherever the majority of matter is concentrated.

When a single graviton carries over a certain number of building blocks of matter beyond a certain threshold at one time, the repulsive energy lines around that single graviton disappear and the surrounding single gravitons move towards that single graviton at the speed of light, carrying any matter structures that appear on them until they collide because there are no repulsive energy lines to keep them apart. The collision creates an explosion of single gravitons and the matter they carry. They take on a different dimension and leave our universe space undetected at a speed greater than the speed of light without interacting with matter or with the fabric of our universe.

The area where this explosion has occurred becomes a black hole, an empty space in the fabric of our universe without gravitons or repulsive energy lines. This black hole becomes progressively bigger as long as matter structures continue to fall inside of it by knocking out the single gravitons that form a ring around the black hole. The single gravitons that form a ring around the black hole have very short repulsive energy lines between them and the black hole, and once matter structures appear on these gravitons those repulsive energy lines disappear and the gravitons fall into the black hole along with the matter structures they carry, creating a bigger hole in the fabric of our universe. A single photon appearing on the last graviton that forms a ring around the black hole has enough energy to cancel the short repulsive energy line between that last graviton and the emptiness of the black hole.

Our Sun has enough building blocks of matter to create a hole in the fabric of our universe that is less than half a mile in diameter. If an empty gap (no repulsive energy between the single gravitons) were created in the center of the Sun by its own gravity force that would squeeze the Sun's mass toward its center. Once there was an empty gap in the fabric of our universe in the center of our Sun, a chain reaction would start. A greater amount of matter falling into this empty hole would make the hole bigger and our Sun would disappear in a matter of a few seconds to become a black hole, less than half a mile in diameter. The size of this black hole would equal the combined sum of the sizes of all the short and weak gaps the Sun contained when it was alive (the short gaps created because of the presence of its own mass).

The distance between the surface of this black hole and the rest of the Solar System would remain the same as that of the center point of the previous Sun, giving this black hole the exact gravity force of our Sun in relationship to the rest of our Solar

System. However, once any matter structure neared the surface of this black hole, the gravity force would increase the closer it came to the surface. Unlike our Sun, which has some kind of surface about half a million miles away from its center point, a black hole does not have a hard surface to stop any matter structure from falling down further.

The surface of the Sun is made from hot compressed gas and plasma (plasma is composed of naked atomic nuclei without electrons). It is in a gaseous state because the high temperature at the Sun's surface (in millions of degrees) causes all solid matter to become gas. On the surface of the Sun the gas and plasma is compressed because the gravity generated by the Sun is tremendous: about thirty times stronger than Earth's gravity force. If you weigh two hundred pounds on the surface of Earth, you will weigh about six thousand pounds on the surface of the Sun.

Inside the core of the Sun, which is about the size of Earth, this hot gas and plasma combine to become a hot, compressed solid liquid due to the tremendous pressure deep within the sun.

At about half a million miles away from this black hole the gravitational force will be the same as it was on the surface of our Sun, but matter structures will keep falling toward the black hole with increasing speed and without stopping. By the time these matter structures get close to the surface of the black hole, their speed approaches the speed of light and the strong gravity force tears the matter apart. Parts of each matter structure that are closer to the black hole even by a fraction of an inch will face a stronger gravitational force than the parts behind them. Eventually, each matter structure will consist only of the building blocks of matter with no connection between them, and the building blocks of matter will become photons by the time they hit the black hole surface, which consists of single gravitons that form an envelope or a ring around this empty hole.

Because matter is torn apart as it approaches the black hole and turns into photons, only photons can reach the surface of the black hole. Once they fall inside they will explode (or shrink) to become a different dimension and will leave our universe at a speed above the speed of light, which for us is an infinite speed because we do not have anything to which to relate it. This different dimension could be at any place and time of the infinite universe, and even though the infinite universe is size-less, its space will collapse to zero because this different dimension has an infinite speed. This is just like the saying that Earth became smaller subsequent to the passenger jet era. Simply because our speed increased, the space of Earth became smaller without its shrinking in size. Earth is about the same size it has been for billions of years, but because our speed has increased, Earth's space seems to have gotten smaller.

Now, imagine that at infinite speed you can be at any point in the infinite universe in no time, that space itself collapses just as if there were no space. Furthermore, time itself collapses and you can be in more than one point in space at one time. On the other hand, if there are no points in the empty space of the infinite universe, there is nothing to which this different dimension can relate its speed, so its speed becomes zero. Likewise, if there is nothing to which it can relate its size, in the emptiness of the infinite universe this different dimension becomes size-less. So, this size-less, speed-less and timeless different dimension could be at any imaginary point in the infinite universe. It could explode or split into many pieces and start its own universe, just like our own universe or a different one, with its own space-time and laws of physics.

When God looks at our universe from the outside, from the emptiness of the infinite universe that surrounds us, He can see a glass bowl with photons expanding inside of it at the speed of light toward the edges as matter structures follow them at different speeds. We can only imagine what our universe might look like from the outside, since there are no photons leaving our universe space, no light to see, no gravitons, and therefore, no feelings of gravity. The matter in our universe cannot survive outside of its boundaries; only our imagination can. This makes our imagination the most powerful force in our universe.

Now, let's return to the topic of the black holes inside our own little universe. These black holes resemble the emptiness of the infinite universe beyond the boundaries of our little universe. We know that black holes inside our universe get bigger as long as more photons are falling inside them. These black holes become so-called "White Holes", creating and putting out matter-energy of a different dimension into the infinite universe. However, if black holes stop receiving matter structures, they become smaller and smaller and can disappear, depending on their shape and size. This can take anywhere from a fraction of a second to billions of years. The smaller the black hole, the faster it will disappear. Like a self-healing living body, the fabric of our universe tries to fill any short gaps or holes caused by the presence of matter structures or black holes inside our universe. The movement of the fabric of our universe towards short gaps or holes is similar to the way that ocean waves hit and chew an island shore from all directions, which for matter structures feels like gravity force.

As we discussed earlier, matter structures are always in motion relative to the fabric of our universe around them. This motion gives the matter structures the energy to offset the gravitational force of their own presence in the fabric of our universe. On the other hand, black holes are stationary. They don't move relative to the fabric of our universe around them, but rather, they move with the fabric of our universe as a part of it. Therefore, black holes do not have any energy. After all, they are just empty spaces.

The constant bombardment of single gravitons towards these empty spaces from all directions makes the black holes smaller and smaller, causing them to disappear from our universe space in a certain amount of time depending on their size, shape and activity. The constant bombardment of single gravitons towards black holes is a result of the vibration of our universe space. The vibration (or boiling) of our universe space is always caused by the motion of matter structures (including photons) which creates imbalance between the negative and positive forces in the fabric of our universe. A movement of any matter structure anywhere in our universe will cause our universe space as a whole to immediately vibrate.

The disappearance of a black hole from our universe space is accompanied by a burst of gamma rays, as some of the single gravitons that form an envelope around the black hole collide among themselves to create photons at their shortest possible wavelength, just as it happened in the Big Bang, but on a much smaller scale.

Black holes are usually a perfectly round ball shape, but in the event of collision with other black holes they can merge to become the shape of a cylinder, a flat disc, or even a crack in the fabric of our universe many miles long. These newly created black holes with odd shapes will have much shorter lifespans since they have more surface area relative to a round, ball-shaped black hole.

Black holes do not wobble and do not circle any other object, not even around a larger black hole or a matter object with a stronger gravity force. This is because black holes are only empty holes in our universe space and they move as the fabric of our universe moves, which is in a straight motion toward any object that creates a gravity field by way of its presence. Therefore, the motion of a black hole is also straight as it moves toward any object with a stronger gravity force. The straight motion of a black hole of any size toward any matter object with a stronger gravity field than that black hole will result in a collision.

A collision of a black hole with a matter object will result in the annihilation of that matter object within a few seconds, causing the matter object to disappear from our universe space. This will create a larger black hole with a stronger gravity force that is equal to the gravity force of the original black hole in addition to the gravity force of the matter object that has been swallowed.

Collision of a black hole of any size with any matter object also creates a huge amount of photons within a few seconds, shooting outward in all directions away from the collision point. The more massive the matter object, the more photons it creates in the collision. These photons come out mostly at the shortest wavelength possible as gamma rays. These photons originate from close to the black hole surface, but not from the surface itself or from the inside of the black hole, because no photons can escape these spaces.

Collision of black holes with other black holes is silent. Since it does not involve matter or photons, but only gravitons, there are no visual shockwaves, no fireworks and no lights. Therefore, the gravity waves change speed and direction in the vicinity of the collision and change the structure of our universe as a whole, since everything in our universe is connected by the motion of the fabric of our universe, making our universe a living body.

Collision of two black holes creates just one black hole the size of the larger of the two. This is simply because we are putting two empty spaces together and we can fit an infinite number of smaller empty spaces into a larger empty space. Since there is no volume to increase, the size of the black hole does not increase and therefore the gravity force of this black hole does not increase either.

In the early stages of our universe, when it was between one to four billion years old, many black holes were created. Since our universe matter occupied far less space than it does today and was cooler than it was during the first billion years, it frequently created dense pockets of matter that collapsed on themselves to become black holes. As time went forward and our universe became older, this large number of black holes collided among themselves to create fewer black holes.

The volume, or the size, of a black hole inside our universe can only be measured from the outside of the black hole and not from inside, unlike our universe, which can be measured only from the inside and not from the outside. The space inside of a black hole is infinite in size, there are no points to refer to, no gravity to feel, and no lights to see. Therefore, you cannot see or feel the boundaries and you cannot see or feel any other object nearby. There is no connection, there is no awareness, and there is no influence of any kind from any nearby object.

Black holes do not spin because a black hole is emptiness, and therefore, there is nothing to spin. The emptiness of a black hole inside of our universe creates a large hole in the fabric of our universe, and as with any hole or short gap in the fabric of our universe, it creates a gravitational field. Any matter object, be it a dust and gas cloud, a sun, or a planet, will fall toward any other object with a stronger gravitational force. Only energy in the form of circular motion can stop those matter objects from falling down further towards the black hole. Because of this, if there is any matter around any black hole you may see those matter objects circling that black hole, giving them the energy to offset the strong inward gravity force generated by the black hole, and keeping them in orbit around it. The lower the orbit, the faster the spin creating friction and heat among those matter objects.

At the surface of a black hole, where the fabric of our universe forms an invisible ring around it, the gravity force is equal to the speed of light, and this means that only photons, which always travel at the speed of light, can circle the black hole very close to its surface without the ability to escape. The spin of all these matter structures and photons around the black hole gives the illusion that the black hole is spinning, but it is only an illusion. Black holes do not spin. Emptiness does not spin, because there is nothing to spin. The eye of a hurricane here on Earth resembles a black hole inside our universe because the wind around the hurricane eye is stronger as it gets closer to the eye and inside the eye there is no wind.

"Naked black holes" are black holes with no matter structures around them. They do not create any heat or a magnetic field around themselves, since only matter structures create heat and magnetic fields due to their spin and motion. Therefore, it is impossible to detect naked black holes by vision, heat or magnetic fields. However, it is still possible to detect naked black holes by gravitational lensing or as cold spots.

There is no such thing as time inside of a black hole. Time is the relative position of matter structures, and there is no matter inside of a black hole. You cannot put a watch or a clock inside of a black hole, since they are made from matter and will explode like all matter in the emptiness of the black hole to then become a different dimension.

The temperature inside of a black hole is zero. This is a state of absolute zero or, more correctly, a temperature-less state, since there is no such thing as temperature or heat inside a black hole because of its lack of matter and photons. All matter, including photons, explodes inside of a black hole, thus becoming a different dimension with different properties. Since there are no photons inside any black hole, there is no light and the black hole is pitch black, darker than the inside of a deep mine sealed with old lead. There are no photons at any wavelength inside the black hole to give warmth or light. You cannot put a flashlight or a thermometer inside of a black hole since these tools are made from our universe matter and like any matter will lose their identity, or in other words, their information unit.

All information breaks down inside of black holes and gets lost without the possibility of retrieval. Since all information is provided by the relative positions of matter structures, and none of our universe matter structures can survive inside of a black hole, everything becomes either infinite or zero: speed, mass, space and dimensions. All the laws of our universe: gravity, energy and space-time breaks down and disappears inside any black hole. Einstein's great theories of relativity and quantum mechanics also breaks down inside any black hole, since we are dealing with much smaller particles than photons, if any, that behave in a completely different manner than anything we know. Those particles (or universes) inside the black hole do not connect or interact with any other particles as there are no forces of attraction of repulsion between them. However, although it is very unlikely, they can still connect or interact with other particles (or universes) by random collision if (and only if) they share the same dimensions of space.

Everything is possible inside of a black hole, but you will never find out because black holes do not have any space to deliver any information. A space is a medium to deliver information and to deliver objects. Earth has different spaces, like hard surfaces for walking and driving, waterways for ships and submarines, and air for flying. Our universe has the web of single gravitons as a space to deliver objects within its boundaries. However, black holes do not have any space, and therefore, we cannot predict how objects move or behave inside of them. Only the human imagination can wander inside the infinite emptiness of a black hole, since it is truly God's territory.

Our universe is a bubble of life. It created, contains, and protects its matter structures from the infinitely harsh and strange conditions of the black hole surrounding our universe. Within this universe bubble, Earth is a bubble of life. Earth created, contains, and protects its life forms from the harsh and extreme conditions of our own universe.

Here is some food for thought: a city with one million people (which can also be described as spaces) can be divided up to one million times to find individual humans. If you divide more than a million it is meaningless: there is no half or quarter human. A room full of air can be divided up to the number of air molecules in the room, if you use a bigger number it's not an air molecule. Our universe's space can be divided up to 10^{2000} (or a one with two thousand zeros behind it, it is a huge number but it's not an infinite number) which is the number of single gravitons that create our universe's space, if you use a bigger number it's not our universe's space.

On the other hand, the emptiness inside a black hole can be divided an infinite number of times into smaller and smaller sizes and you'll still wind up with emptiness. All this means is that a size relative to a black hole's emptiness is meaningless; it could be infinitely larger or infinitely smaller at the same time. A black hole is the realm where the very large and the very small is the same, and why relativity breaks down, the place where you must get lost, the place of no return.

11. Our Multi-Dimensional Universe

Before the Big Bang our universe was one-dimensional, a size-less point in an empty space. After the Big Bang, this single point split into about one with two thousand zeros behind it of points (gravitons) and created the fabric of our universe.

Some of these single points paired up to create the building blocks of matter, a two-dimensional matter (like two dots connected by one line of energy). The energy line between each of these two dots fluctuates in length, depending on its energy and on outside forces. It can get longer or shorter, controlling the spin of the building blocks of matter. The longer the energy line, the faster the spin. This motion is the third dimension.

So, a one-dimensional point in space becomes two-dimensional just by pairing up with another point, and their motion, as the points change distance and position from each other, creates the third dimension. This is an example of only two out of a huge number of building blocks of matter, and every building block of matter has its own space-time. Some of the building blocks of matter group together to form matter structures, and that means that all matter structures are multidimensional, since they are all made from a huge number of single points.

In theory, the number of possibilities to create different dimensions from the combinations and connections of all those single points is over one with two thousand zeros behind it (10^{2000}) in our universe alone. It is a huge number, but it is not infinite, because the number of points of which our universe is made is finite. This means that if there is an infinite number of points, there will be an infinite number of dimensions, and that is the case of the infinite universe surrounding our universe which contains an infinite number of universes.

It is hard for us to comprehend an empty space because our universe space is made from a huge number of points. So, this is our space, our universe space, and the connection of all those points gives us the size and the volume of space. The easiest way to understand volume is by using three dimensions of space. Four walls, a floor and a ceiling together give us three dimensions of space. Up, down and sideways motions, and motion itself, give us the fourth dimension of time. Without the fourth dimension (motion) any multi-dimensional space becomes one-dimensional because you cannot realize it without motions such as up, down and sideways.

So how can we figure out the volume of a multidimensional tree, with all of its branches and leaves that have different sizes and shapes? We can only determine the volume through a long list of calculations, or simply by taking the whole tree and immersing it in a pool of water to figure out the true volume based on the volume of water that rises.

The easiest way for us to figure out the volume of any multidimensional matter structure is to compress it to three dimensions of space. However, all matter structures in our universe, including the fabric of our universe itself, are made from a huge number of points. These points, gravitons, and their connection to other gravitons, can lead to a huge number of dimensions in our universe alone. So our universe is a multidimensional universe.

12. Photons and our Universe's Missing Mass

"Nature repeats itself in many strange and amazing ways. All the living organisms on Earth were created from bacteria (the smallest living organisms). Bacteria outnumber and outweigh all other living organisms by many folds. All the matter in our universe was created from photons (the smallest units of matter). The photons outnumber and outweigh all the matter in our universe by many folds." - God

All the matter in our universe was created in the Big Bang. The Big Bang was the explosion or the splitting of one-dimensional matter into a huge number of gravitons. At the central point and time of the Big Bang, where the gravitons were denser, some of them paired up with no distance between them and created matter in the form of a huge number of photons shooting outward in all directions at the speed of light.

Now, all those photons were created in the center of the Big Bang in a short time period and in a tiny space, which created very high-energy shortwave photons. The collision among the huge number of photons in a small space created very small subatomic particles, mostly in the form of neutrinos which are made from at least two pairs of gravitons. A single photon is made from two gravitons with no distance between them. Head-on collision between two photons will create a single neutrino, which has two pairs of gravitons with a very small distance between the two gravitons in each pair. These two gravitons, which are the building blocks of all matter, spin around each other connected by one line of energy. The longer the distance between them, the faster they spin around each other.

The spin of these building blocks of matter likewise creates spin in the subatomic particles they create through their grouping with other building blocks of matter. The spin of subatomic particles creates magnetism. Because of the spin they can annihilate, repel or attract other subatomic particles nearby depending on the direction of the spin and also on the number of building blocks of matter they contain. The attracting and repelling forces between subatomic particles are the cause of their huge and diverse numbers. These forces led to the creation of bigger subatomic particles, and later on to the creation of hydrogen atoms, all in a span of just a few thousand years after the Big Bang.

Only about fifty percent of all photons that were created at the point and time of the Big Bang collided to create subatomic particles. However, they created about the same number of subatomic particles spinning clockwise (matter) as subatomic particles spinning counterclockwise (antimatter), with a slight advantage of about one percent for the subatomic particles spinning clockwise (matter) due to the angle of collision of photons created in the Big Bang. Collision of equal quantities of matter and antimatter resulted in total annihilation of both the matter and antimatter involved in the collision. This matter and antimatter transformed back into photons, and some of these photons collided again to produce subatomic particles. This process continued for the first few years after the Big Bang until most of the antimatter disappeared due to its initial disadvantage of one percent. After a few million years, as our universe matter expanded and occupied more space, it cooled down, and at this point consisted of about ninety percent photons and ten percent matter in the form of clouds of hydrogen atoms that formed a huge number of galaxies. Some of the matter in the center of the galaxies collapsed on itself as it created enormous gravitational forces to form black holes. Some of the left over clouds of gas formed a huge number of suns.

The sun, a huge body of matter made mostly of hydrogen atoms, creates an enormous gravity field that squeezes the atoms in its core to create heavier atoms by fusing the hydrogen atoms together through pressure. The byproduct of this fusion is a huge number of photons shooting outward in all directions. A sun of average size loses a few thousand tons of matter every second as it transforms matter into energy in the form of photons. There are billions and billions of suns in our universe and that means we are losing many trillions of tons of matter every second that is transforming into photons again. Some of these photons transform back to matter again once they hit a matter object, adding their mass and energy in the form of heat and motion to the body of the matter. However, most of those photons are lost in the huge void between the bodies of matter.

Today, fifteen billion years after the Big Bang, our universe matter (mass) consists of about ninety-five percent photons and five percent regular matter, compared to a ratio of about ninety percent photons and ten percent regular matter only a few million years after the Big Bang. So, in about fifteen billion years we lost about fifty percent of the matter in our universe. Some of this lost matter transformed again into photons through collision of matter structures among themselves, as it happened mostly but not entirely in suns, and some of the matter was swallowed by black holes to become a different dimension and leave our universe space without a trace.

There is a very important law in physics which basically says: matter-energy never disappears, it only changes form. Therefore, the loss of matter means that photons account for the missing mass of our universe, the black energy that cannot be seen and is only known to us because of its gravitational pull. It is true that we cannot see photons unless they hit our eyes or we feel the heat they cause by collision with regular matter. But every cubic inch of empty space up to fifteen billion light-years away from the point of the Big Bang in our ball-shaped universe contains a huge number of photons of different wavelengths, all traveling at the speed of light and in different directions. The large voids between galaxies far away from any sun still have enough photons in them to keep all matter objects at a temperature of three degrees above absolute zero through collisions of these photons with any matter object.

All matter objects (except photons) produce photons as a byproduct of the spinning and the collisions among the huge number of the building blocks of matter contained in each matter object. The longer the distance between each two gravitons that form the building blocks of matter, the faster their spin around each other. The faster their spin, the more collisions they have with other building blocks of matter and the more photons they produce. The more photons a matter object produces, the more mass it loses and the colder it gets as the two gravitons get closer to each other and have less spin. This will result in fewer collisions and reduced production of photons.

So without receiving outside photons by collision, matter objects will steadily get colder all the way down to absolute zero and disappear (decay) to become photons in a matter of a few thousand to a few billion years depending on the distance between the two gravitons. The longer the distance, the faster they will decay, and the shorter the distance, the more slowly they will decay. If matter objects receive more photons than they produce, they become hotter and more massive. If matter objects emit the same number of photons as they receive from the outside, there is no change in their temperature or their mass.

For example, heavy atoms are made from many subatomic particles, and some of these subatomic particles are made from a huge number of graviton pairs. These atoms decay much faster than lighter atoms to become lighter atoms themselves, or they break down into many subatomic particles, producing many photons in the process. The electron is made from six building blocks of matter. The six building blocks of matter that make up a single electron have a very short distance between the two gravitons of which each building block of matter is made, and that makes the chances very small that those six building blocks of matter will collide among themselves to produce photons. Therefore, they can disappear or decay to become photons only after many billions of years (lifespan). The photon, on the other hand, is made only from a single block of matter with no distance between the two gravitons, and therefore, there is no spin and no collision between the two gravitons. A single photon does not produce any more photons, so paradoxically, the temperature of a photon is absolute zero (dark matter).

No spin and no change in the relative position of the two gravitons of which the photon is made also means it does not have internal motion, which is the fourth dimension, as a result of which time does not flow for the photon. It is timeless. It always stays the same and therefore does not have a lifespan and will never decay. Because there is no spin between the photon's two gravitons, there is no anti-photon. No spin between the two gravitons also means the photon does not create a magnetic field. Therefore, there is no attraction between different photons allowing them to form a bigger body of matter. Photons do not group or pair with other photons. They always stay single, unlike the building blocks of matter that always group with other building blocks of matter. There are no single building blocks of matter. If a building block of matter gets separated from the group it becomes a photon right away. On the other hand, there are no photons connecting to the building blocks of matter. If a photon is created inside a matter structure it leaves it right away, and if it collides with matter it becomes a building block of matter (a pair of gravitons with the distance between them). Finally, because the photon is made from two gravitons with no distance between them it is a dense matter object. It is heavier and therefore it creates a stronger gravitational field relative to a single block of matter.

Photons do obey the law of gravity. They do ride on the single gravitons for the duration they appear on them, and they do create a gravitational field by way of their presence on the fabric of our universe.

All matter structures create gravitational fields by forming weak (short) gaps in the fabric of our universe. The more matter is present and the more concentrated it is, the more short gaps it creates in the fabric of our universe, and the weaker those gaps are. The fabric of our universe closes those weak gaps with the movement of its gravitons towards the central gravitational point of the matter structure. This is the gravity force, as those single gravitons carry the matter for the duration the matter appears on them towards the central point of the weakest gap the matter creates by way of its presence.

The photon, which is made from a single block of matter, will create a gap with the least possible weakness in the fabric of our universe, which will create the weakest gravity field possible and will be hardly noticeable close to a matter structure of any size. Additionally, we cannot compare an empty space with a space that is full of photons in terms of its gravitational field, since all the empty spaces around us are full of photons. But the photons, because of their huge numbers, create a huge number of gaps with the least possible weakness in the fabric of our universe, and this is profound in the large empty spaces in and in between the galaxies, and in the empty spaces in the front of the galaxies towards the edges of our universe.

The gravity field, or the combined mass the photons create in a single cubic light-year, is approximately equal to the mass or the gravity field of our entire Solar System, and there are trillions and trillions of cubic light-years in our universe that have only photons in them. The gravitational force of those photons is pulling and accelerating the movement of galaxies at the frontal edges of our universe closer towards the edges of our universe, because most of the photons are ahead of matter as they travel at the speed of light from the point of the Big Bang towards the outer edges of our universe.

A gravitational force always pushes the lighter mass towards the heavier mass and makes heavy objects spin faster than lighter objects. Gravity force does weaken over distance, but never disappears, even over distances of billions of light-years.

Our universe mass consists of about ninety-five percent photons and about five percent matter, and therefore, the larger mass of photons creates a stronger gravity force. This can be witnessed in the fast spin of galaxies and clusters of galaxies, which indicate that they are much heavier than what they seem.

Therefore, the photons are the so-called "black energy" that accelerates the motion of outer galaxies toward the edges of our universe by way of their gravitational pull.

Also, the photons are the so-called "black matter" that prevents galaxies from tearing apart despite their fast spin.

13. The Fate of Our Universe

God set a timetable to end our universe in its present form by enforcing the laws of physics he imposed on our universe. This is the lifespan imposed on any object that has internal motion.

About fifteen billion years ago our universe started as the Big Bang, the splitting of matter-energy of a different dimension into a huge numbers of gravitons. Those gravitons created the fabric of our universe which stretched thirty billion light-years in all directions, away from the point of the Big Bang, and defined our universe space and boundaries. Beyond the boundaries of our universe space is a black hole of endless size, the emptiness of the infinitely large universe around our infinitely small universe.

At the point and time of the Big Bang where the gravitons were denser, they created the building blocks of matter by pairing up with other gravitons through collision. The building blocks of matter are always in motion. Motion gives them the energy to withstand the gravity force that the fabric of our universe imposes on any matter structure. Therefore, their motion is always above zero relative to the fabric of our universe, up to a maximum of the speed of light, which is the speed limit likewise imposed by the fabric of our universe on any matter structure.

As time moved forward from the Big Bang, the building blocks of matter grouped up to form all the atoms, planets, suns and galaxies in our universe, with their motion continuing away from the point of the Big Bang and toward the edges of our universe.

As our universe's matter expands, the distance between galaxies increases, and they become colder while their speed accelerates toward the edges of our universe. This means that as our universe gets older the distance between the galaxies will increase, the size of each galaxy will increase, the overall temperature of each galaxy will decrease and with it their speed outward will increase. This will happen simply because all matter structures are made from the building blocks of matter, and as the distance between each two gravitons that make up the building blocks of all matter decreases, the matter will have a higher speed through space and will produce less heat in the form of photons. The photons that come out of that matter will be of longer wavelengths (lower energy).

Once these galaxies reach the distance of about twenty billion light-years away from the point of the Big Bang, the distance between the galaxies will become much larger and each galaxy will increase in size, become colder and move faster. By this time, gravity force will also increase, as these galaxies get closer to the edges of our universe. The gravity waves will become stronger and faster the closer they are to the black hole around our universe, increasing the speed of the galaxies toward the giant black hole around our universe.

Around fifteen billion years from now all the galaxies will tear apart and there won't be any galaxy formations. Our universe temperature will be only a fraction of one degree above absolute zero. All matter structures will approach the speed of light as they head toward the giant black hole (the emptiness of the infinite universe around our universe) without the ability to escape, by creating circular motion, since this black hole is all around them.

Once the matter structures approach the distance of a few thousand miles away from this black hole they will tear down to their building blocks, and by the time they hit the black hole, they will be photons traveling at the speed of light. Once they leave our universe space they will explode to become a different dimension, meaning they can be anywhere in the infinite universe in no time and can even start their own universe. Each photon can split to create its own universes.

The end of our universe will also satisfy our laws of physics: the law of matter-energy, the law of gravity and the law of time.

The law of matter-energy states: matter becomes energy only by motion, so the motions of matter structures are always away from their center, away from the point of their creation (which is the Big Bang), and toward the edges of our universe, toward the black hole around our universe.

The law of gravity states: gravity waves will carry any matter structure with them toward the shorter gaps or holes that create a stronger gravity field. The black hole around our universe creates the biggest gravity field of them all.

The law of time states: If there is a beginning, there must be an end, and the end will resemble the beginning. The ultimate beginning of all matter structures is that they were created from photons and the photons were created from a different dimension. In the end, all matter structures will become photons before the photons themselves become a different dimension.

Once our universe expels all its matter it will become a one-dimensional object with no internal motion, floating motionlessly in the unlimited emptiness of the infinite universe for an undetermined amount of time.

Later on, it could collide with another object, starting a new universe. Or it could disappear to nowhere. Only God knows!

So get ready for the judgment day. It will happen in about fifteen billion years from now, and there is no escape.

14. Lifespan

All matter structures have to have lifespan. This justifies the law of time which says that if there is a beginning, there must be an end. Time itself is a byproduct of the motion of matter structures that change their position relative to other matter structures, causing time to tick forward. The period of time for which a matter structure stays more or less in the same form is the **LIFESPAN** of that matter structure.

As you will see later on in this chapter, lifespan depends on the force and the number of interactions of a body of matter with the forces of energy and gravity. Lifespan does not depend on time itself, since time is only the byproduct of the movement of matter structures caused by either energy or gravitational forces. Lifespan can also be applied to energy, because energy is matter in motion. So, energy has lifespan just like matter structures, and it is dependent on the matter structure that carries it.

Let's take the photon for example. The photon is the smallest unit of matter. One simple atom of hydrogen has more than a million times the mass of one photon, but because the photon always travels at the speed of light it has a lot of energy for such a small amount of matter. The photon is made from two gravitons, but unlike the building blocks of matter, the two gravitons of which the photon is made have no distance between them and therefore no spin. The photon does not have any moving parts and does not decay like regular matter. So, a photon that was born fifteen billion years ago in the Big Bang looks just like a photon that was born eight minutes ago in our Sun. It is as if it is timeless, and this is true, because we know that time stops at the speed of light.

Our Sun produces a huge amount of photons and their lifespan depends only on the distance they travel before they hit an object. At that moment the photon disappears, adding its mass to that object and its energy in the form of heat, higher motion, or faster spin to the matter object. In the case of a photon coming from the Sun (its birthplace) and hitting Earth, its lifespan is about eight minutes, because that is the time it takes to cover the distance from our Sun to Earth. The energy of the photon will only materialize when it hits atoms in the Earth, and that will be the end for this photon. But most of the photons made by the Sun travel in Space for billions of years unchanged before they hit an object and die, or to put it more correctly, change form, so the lifespan for these photons can be billions of years. The photons have only one interaction before they disappear and become something else. The lifespan of all photons and the matter structures in our universe cannot be longer then the lifespan of our universe itself, which also obeys the laws of time saying that if there is a beginning, there must be an end. Our universe was born in the Big Bang, coming from a different dimension, and it will eventually return to this different dimension.

Everything has a lifespan. Nothing lives forever in the same form. From the very small photons to the very big objects, this is true for our whole universe and everything in it. All energy and all matter changes form constantly, which is another hallmark of time. The same is true for manmade products like airplanes and cars. Manufacturers know the lifespans of their products based on average use (the length of time before things start to fall apart and need to be replaced or repaired) and they give a warranty that usually expires before those products need to be replaced or repaired.

The same is true for nature-made genes: they all have a lifespan. Genes are made from complex organic molecules that are always in motion creating new molecules to replicate themselves or to create a substance (protein). There is only a certain number of times they can divide to make new cells and repair themselves before they change form. That certain number of times is the lifespan of each gene. For example, the genes that build the human body have a lifespan of up to one hundred and thirty years, while atoms have lifespans of up to billions of years that can end only by collision or decay. Genes act more like the molecules of a rechargeable battery, can last only for a certain amount of time depending on their use and the conditions of use, and can only be charged a limited amount of times.

The rechargeable battery is made from molecules of certain atoms that consume energy in the form of electrons. These extra electrons, coming in from an outside source, move the molecules inside the battery to different positions by disconnecting molecules that share their outer layers of electrons with other molecules. Instead, they use the extra electrons they receive. All this motion creates heat. The extra energy is stored in the form of extra electrons. When the battery produces electricity, the whole process repeats itself in reverse. The molecules lose electrons and then get attached to other molecules by sharing their outer layers of electrons. The motion of all those molecules, atoms and electrons also produces heat. (The more efficient the battery, the more heat it will produce.)

There are billions of molecules in a small battery. Every molecule has a few atoms and those atoms have at least two loose electrons depending on the substance of which the battery is made. So, we are talking about a huge number of molecules, atoms and electrons. But with the movement of the molecules every time the battery is charged or used, some of these molecules change form. Their molecular structure changes, either by the changing positions of the atoms in the molecule or by the loss of some atoms or connection to extra atoms. Either from an external or internal source, the molecules lose their ability to store and produce electrons and the battery no longer functions like a new battery. After a certain number of uses, none of the molecules will be functioning and the battery dies. So the battery lifespan depends on the number of interactions, and not on time.

The same is true for genes. Genes are organic molecule that can produce a substance, replicate or repair themselves only a certain number of times before the gene molecules attach more atoms or lose some of their atoms (either from an external or internal source). The atoms comprising the molecule then become structured in the wrong position/order, upon which the gene loses its previous identity and the gene molecules do not create their protein at all or the gene molecules make the wrong copy of themselves. Again, just as with the battery molecules, all this does not happen all at once. There are thousands of different genes in the human body, and this process takes time. It accelerates when there are more interactions of the gene dividing to make new cells.

The human body is a sophisticated organic machine built and designed by the genes that pass from one generation to the next. These genes have a self-repairing mechanism, but repair becomes less efficacious over time. A smoker will have a shorter lifespan because he has more interactions with pollution and his body needs to clean and repair itself more often. Overeating or eating the wrong food causes the body to have more damage and stress, and that will shorten the lifespan. Too much exercise or hard labor will shorten the lifespan. Sickness or disease will cause a shorter lifespan because the body will have more interactions with harmful substances or medicine and will try to repair itself more often.

On the other hand, sleeping extends the lifespan because the body relaxes and has fewer interactions. All this does not mean that if you do not eat or exercise you will live longer. Everything has its proportion and you have to do what you are designed for by your genes.

If you take two new electric poles and put one near the ocean in humid, hot conditions and take the other one and put it in a location with a pleasant, dry climate and come back after ten years, just by looking at these two different poles you will see a big difference between them. The pole near the ocean will look much older, because it has had more interactions with water and salt. It will disintegrate and disappear long before the other pole, even though they are of the same age. A brick used for building houses has a lifespan of up to one thousand years under normal weather conditions, but if it were to get hit by a sledge hammer a few times it would become sand and dust. So, lifespan really depends on the number of interactions and the force of those interactions, which can change the quantity of a given matter structure or its shape, and by doing so change its identity entirely, changing the message it represents.

Nobody can live forever, and nobody dies from old age per se. There is always a physical reason for dying, for reaching the end of the lifespan.

15. Nothing Is the Same

Every time I go back to my old hometown by the Sea of Galilee, which is a medium sized lake in Israel, it looks to me just the same as I remember it looked fifty years ago when I was growing up there. It gives me a sense that I am young again and have gone back in time. But is it really the same? The lake water is not the same water as it was fifty years ago. The old water has long gone through evaporation, human consumption and by flowing away to the Dead Sea via the Jordan River. The fish in the lake are different from the fish of fifty years ago, and the size and shape of the lake has also changed because the water level fluctuates all the time. My parents died, I lost a few friends and gained a few new ones since then. My thinking has changed and I do not look the same as I did fifty years ago.

All these changes are hallmarks of time that say all matter structures must have motion through which they change their relative positions, causing time to tick forward. Matter structures can be in a solid, liquid or gaseous state, and their motions cause interactions among the huge number of matter structures of which our universe is comprised to form new matter structures of different shapes and sizes that occupy different spaces. Time stops with the disappearance of an old structure and restarts, ticking onward, when a new structure is created.

Let's take a comet for example. A comet is basically a body of rocks, dust and ice the size of a large mountain, that circles our Sun for billions of years until it suddenly hits Earth. Once it hits Earth, time stops for this comet, and it becomes something else. It becomes integrated as part of Earth, losing its shape, its uniqueness and its identity. Time stops for this comet as a comet and restarts for it as an integrated part of Earth. It is not a comet anymore.

The same thing happens to Earth in this scenario. It is not the same as it used to be. Time stops for Earth and restarts again after it gets hit by the comet. Earth's mass increases by a few billion tons, its water level increases, its shape changes where it gets hit, and even Earth's motion and spin change by a few seconds a day. Time stops for Earth when it gets hit by the comet and restarts again for a newly improved, heavier and slightly different Earth.

The comet becomes something else internally, because it changes space and different spaces have different conditions. Time ticks forward as this matter structure changes position and by doing so changes form as the condition changes at the new location or position in space. The same is true for Earth. Earth is a unique place. It is one of a kind, a product of its own special surrounding conditions, along with its size, the Moon's size and the Sun's size, which also determines how long the Sun will shine. Another factor is the distance of Earth from other heavenly bodies including the Sun, Moon and other planets in our Solar System, along with the stuff of which our Solar System is made, including comets and asteroids. Because of all of these variable conditions, we can say with absolute certainty that there is no other planet exactly like Earth in our whole universe which contains many billions of planets. Some of them will be similar to Earth, but none will be exactly the same. Therefore, even if some of those planets that resemble Earth in its uniqueness also have life forms on them, these life forms won't be the same because they will have been subject to different conditions throughout their evolution.

When you go to a small or large forest with millions of trees, you will never see two trees that are exactly alike. Even two trees that came from the same tree or seeds and grew up at the same time and at nearby locations will never have the same height, the same number of leaves, the same position and color of leaves, or the same shape and size of leaves and branches as any other tree. The reason is that although they come from the same parent tree and their genetic materials are practically identical, they do not occupy the same exact spot at the same time, and different spots or locations have different conditions.

Let's take the wind for example. Wind that comes from the west will hit the first tree on the west side and will then disperse and lose some of its energy before it hits the next tree. Different ground locations will also have different amounts of nutrients, and water that affect tree growth and shape. Exposure to sunlight, temperature and gravity force are also factors that vary at different locations. A tree that grows on a mountain and the same kind of tree growing in the valley below will be exposed to totally different conditions. The temperature will be a few degrees warmer overall in the valley, and the wind will have less force in the valley compared to higher ground. The ground in the valley will be more nutritious and moist, since rainwater runs downward to the valley from the mountain, taking with it some nutrients. The Sun's rays will be brighter on the mountain since they will have less atmosphere to go through, and hence the tree will have more ultraviolet light. Even gravity will be a little bit stronger in the valley because it is closer to the center of Earth then the mountain is. So, all those trees face different conditions because they do not occupy the exact same location at the same time. Different locations have different conditions and different forces, and therefore, the trees respond to those different conditions by having different shapes, colors and sizes, and by doing so become different and unique. They are each one of a kind in our universe.

The same is true for snowflakes. No two snowflakes look exactly the same. Sometimes you may need the help of a microscope to tell the difference, but nevertheless, they are all different from each other. This is so for the simple reason that water vapor in the air goes through a unique time and space to crystalize and become a snowflake. The snowflakes look the way they look because of the collision and the attraction of minute crystals of ice with other crystals of ice to form the snowflakes. They are all different from each other in shape, size and the number of ice crystals in them, because they go through different spaces in the atmosphere and therefore experience different conditions before they hit Earth's surface. Each snowflake goes through a different air pressure, air speed and air temperature with different levels of dust, sunlight and electric charges, making them a lot or a little different from one another.

The same is true for grains of sand. No two are alike. All the trillions and trillions of them differ from each other in size, weight, shape and color. You probably need a microscope to tell the difference, but none are the same, because they all were created at different times or locations and had different interactions with different matter structures or different energy levels.

Even the best microscope won't help you to see the difference between atoms of the same kind. Atoms are the smallest unit of matter that can stay in the same form over a long time period. Just like trees, snowflakes and grains of sand, no two atoms are identical. For example, the oxygen atom has eight neutrons and eight protons in its nucleus and eight electrons circling the nucleus. All these subatomic particles that make up the oxygen atom nucleus are made from thousands of smaller units of subatomic particles, and all those subatomic particles are made from up to thousands of different numbers of the building blocks of matter, the pairs of gravitons that are connected by one line of energy, with the distance between them always changing depending on their speed in our universe space and the gravity force to which they are subjected. As we saw earlier, the distance between the two gravitons determines their spin. The closer they are, the slower the spin, and the farther they are, the faster the spin.

So, as you see, the atom is a complex matter structure made from a huge number of the building blocks of matter, that just like any other matter structure never occupies the same space at the same time, and therefore, is subject to different forces of energy and gravity. Furthermore, not all atoms receive and discharge the same amount of photons. An atom receiving photons will grow in size, become heavier and spin more quickly. An atom discharging photons will shrink in size, become lighter and spin more slowly. So, no two atoms are exactly the same. They differ in size, spin and their overall number of building blocks of matter, which explains why some atoms will decay before others.

In contrast, all photons are identical. A photon is made from a single building block of matter, which is the pair of gravitons with no distance between them. Photons have no spin and no moving parts, and their speed throughout the fabric of our universe is constant. They all have the same exact speed, energy and mass, which makes all photons the same. So, all the matter structures in our universe are unique, one of a kind, and none are the same, except for the photons.

All matter structures in our universe are made from a huge number of smaller units of matter structures, and as they go through a different space-time, their shape, size and motion is altered to become different from other matter structures. The various structures never occupy the same space at the same time, and therefore have different levels of energy and/or mass and different shapes. They become different from each other, they become diverse, and none will be the same.

If you look at a group of people from far away they all look the same, but when you get close to the group of people you can see that they are not the same height, the same color or size, that they all have different voices and different faces, different life experiences, and therefore, different ways of thinking. Even if all humans were created equal, they are sure not to have the same exact conditions growing up, which makes them different. But is this also true for identical twins that grow up in the same house? The answer is: yes. The twins share identical genes that will make them look and behave identically if they grow up under the exact same conditions. However, that is impossible because two people could never occupy the exact space at the same time.

For example, let's take identical twin brothers. We'll name one of the twins Albert, and his brother Isaac. The twins sit down one day to share one bowl of salad made from lettuce, cucumbers, tomatoes and pitted green olives. Having no knowledge or experience that some of the olives could still have the pit in them, they enjoy the salad and chew hard on it until one of the twins (Albert) chews hard on the olive pit and painfully breaks one of his teeth. Now you can tell the difference between these two identical twins that look the same, dress and behave the same. Whenever Albert opens his mouth, you can see a broken tooth and you know it is Albert.

Now you can tell the difference between the identical twins, but the difference does not stop there. Albert will behave differently whenever he sees a bowl of salad, and even just thinking about it will cause his heart rate to increase a little and his blood to pressure go up. He will have minor anxiety attacks and perhaps some nightmares, as compared to his twin brother Isaac, who will not, because he didn't have the same experience. And as the twins get older, their appearance and behavior will become more diverse.

Let's say at age eighteen they have girlfriends that look similar. Since the twins have the same reference, their identical genes will build and position approximately the same cells in their brains, which in turn will give rise to similar behavioral, thinking and reference patterns. Isaac's girlfriend May is more homebound and quiet, and she stays with Isaac in New York City. Amanda, Albert's girlfriend, has similar characteristics as May, but decides to study for four years at the University of Florida in Gainesville and Albert decides to go with her. After four years in Florida, Albert and Amanda come back to New York City, and the difference between the twins is much more profound. Albert looks tanner and a little bit older than his twin brother Isaac. Albert moved to a totally different environment from his twin brother, and he had to work harder to provide living income for himself and his girlfriend Amanda, and also experienced a few hurricane storms while he was in Florida. Even his accent changed a little bit to have the hint of a southern twang.

So, the more diverse the conditions and the longer the time of separation, the more profound the change. It is interesting to know that the same rules that apply to diversity and lifespan are also correct here. The change will be more profound and diverse when there are more interactions and also depends on the force level of each interaction.

As time goes forward nothing is the same when applied to any matter structure, even within the same matter structure, unless you apply energy to reposition the matter structure the way it was before. A matter structure will always change with the passing of time, as it interacts with internal and external matter-energy. Matter structures will change forms, and that also includes the human's body and brain. We perceive these changes as aging, and as our brain changes with the passing of time, so does our thinking, since our thinking is a reflection of certain matter structures (molecules) in our brain, and as these molecules change their order or positions due to interaction among themselves, our thinking will also change.

Many things can alter the matter structures in the brain. This includes drugs, alcohol and blood flow, as well as new information we receive through our senses that turns into electrons and molecules, including memory molecules from the memory cells.

How many times have you said, "I will never change my mind," but you do eventually because the conditions change.

So basically, nothing stays the same because as time goes forward the position of everything changes, and with it the conditions change. Therefore, matter structures have to change to reflect changing conditions, and we also have to change, since we are also made from a huge number of matter structures.

I can hear God laughing in the background saying, "There are a lot of people who believe that their brain is made out of something special, and not matter, and that is very funny."

Nothing is the same also answers the riddle of the chicken and the egg: which came first? The answer is very simple. Everything changed through time and evolution not to be the same. Therefore, today chickens and eggs only resemble the chickens or eggs of millions of years ago. And, many millions of years before that they were totally different.

So, who came first, the chicken or the egg? Since everything has to have some kind of a beginning, we can answer this question using another question: who came first, you or your parents? As you will see the answer won't be the same for everyone, the answer depends on who you ask and how far back in time you want to go. You will answer the last question saying your parents came first, but your children will say that you came first. The chicken will say it came from an egg, and the egg will say it came from a chicken. But how has all this started, how did life on Earth start?... All life forms on Earth are related to one another and they all started from a single gene like molecule that replicated itself under Earth's right conditions. The first few replications were identical, but as the conditions changed, so did the replications, as they are not to be the same which is what evolution is all about, and in a span of 4 billion years created all the diverse and numerous life forms here on Earth. So, you can say that the first gene-like molecule was the egg that developed the second generation, which was the chicken, or you can say that the first gene was the chicken that lay the egg. The answers are not the same, but they are all correct.

Here is another question that the answer to it won't be the same. The answer depends on who you ask and when. The question is: who came first, God or humans?... The believer's obvious answer will be God, of course. The non-believer's answer will be humans came first. The dumb will say that they both came at the same time. And the dumber will say that there is no God or human, there is only animal. All these answers are correct since there is no absolute truth; everything is only relative to the individual human mind. Whatever you believe is correct, is real, and it comes from your mind. The mind has the final say what is the truth, and no two minds are the same.

16. Diversity

It is hard to believe or imagine that everything we see and feel in our universe, all the energy and all the matter with all its different sizes, shapes, and colors, came from a single size-less point in an empty space. You can call it God's seed which split up into a huge number of gravitons that by themselves are points in an empty space and are therefore one-dimensional and identical to each other. Some of those gravitons paired up and started to rotate around each other at different speeds, directions and distances from each other, and the diversity started. Those pairs of gravitons are the building blocks of all matter, including photons.

A photon is also made from a pair of gravitons, but unlike the building blocks of matter, there is no distance between the two gravitons and therefore no spin in either direction, so they are one-dimensional and therefore identical to each other. They do not group up because they are identical, and not diverse. But when the photons are created they have a space between them, and this space is called a wavelength. The shorter the space between the photons, the shorter the wavelength and the more energy it has. Shorter distance means it has more photons per given amount of space. These distances can vary widely. There are over one with forty zeros behind it of different wavelengths. The different wavelengths give us all the different colors we see, which come from identical photons that only differ by the size of the spaces between them. This gives us all the colors of the rainbow as well as other wavelengths we can't see, from the low energy infrared rays to the higher energy ultraviolet rays and ultimately to the highest energy gamma rays.

The building blocks of matter grouped together in many combinations to create the many diverse subatomic particles, which in turn grouped together in many combinations to form the many diverse and different atoms. The diverse atoms then created diverse molecules and this continued up to the level of much larger matter structures. So, diversity is in the roots of our universe, and diversity has increased. From the very small subatomic particles, atoms and molecules to the very large and complex planets, stars and galaxies, no two are the same, and they are all always changing form as time goes forward.

Our universe looks like a piece of art in the making, or like God said, "That's the way it is, so let it be."

We know that matter structures do not stay the same, that they always change shape, speed, mass and location, and as they change location their conditions change and they are subject to different energy levels, such as photons, gravity force and collision with other matter structures, and as the conditions change, matter structures adapt to them by changing their structures.

Changing structure is another hallmark of time. Time will not flow if there is no movement or change in the structure of any unit of matter, from the stars to the atoms. They all obey this rule and they cannot stop their motion. If they stop, they will collapse from their own gravitational pull and will take on different dimensions.

All matter structures change with time. It could be a minute difference, but it is a difference nonetheless. This means that every unit of every matter structure is unique (one of a kind). Therefore, matter structures are diverse. Diversity is a fundamental part of nature, and since nature repeats itself in many different ways, we can see diversity everywhere. The more diversity is present in nature, the better the chances of survival become.

Let's have a few examples from the life forms on Earth. Viruses often change their outer layers, or envelopes, in order to bypass the host's immune system, which is always trying to eliminate harmful invading viruses. Viruses use diversity to deliver their own genetic code, their own identity, intact. The human body is made from a huge number of different types of cells, and the key for cell survival is adaptation to different environmental conditions, or the ability to resist them, be they chemical, biological, weather related, or related to the availability of matter-energy.

However, as all matter structures resist changes using energy, and since the living cell is also a matter structure, it resists change by storing its own energy and keeps itself more or less in the same shape and order despite changing conditions. In other words, it is alive. It has its own energy to counter external energy that could change it. This way the living cell can keep its own structure, its own identity. But its energy is limited against the seemingly infinite outside forces and the flow of time forward, so it tries to bypass this inferiority by duplicating itself through the genes as closely as possible to its original form, and it has a little success, since it always will have some changes as it goes forward in time. This happens for generation after generation as the conditions constantly change, which is what evolution is all about.

So, it is fundamental that matter structures resist change, but they always lose over time to external energy forces (lifespan). The same is true for the living cell, which resists changes but changes nonetheless over time due to its interaction with outside matter-energy to become a little different than before.

We as humans resist changes. We resist diversity, and this resistance manifests itself as an urge that comes from within our brains telling us not to mix with other kinds. This is the main cause for racism. So racism is basically an urge that comes from the subconscious brain, which reflects the resistance of matter structures to change and to become diverse. This urge for racism usually develops when the human being reaches sexual maturity, or in other words, when it is able to copy itself. People try to duplicate themselves by sharing genes only with other people who are similar to them or who have the qualities they like or with which they can identify. They then have offspring that resemble themselves, or are even better than themselves, offspring that will have a better chance of survival and by surviving will pass on their own genes. In other words, people do not want to waste their energy on offspring that do not resemble them or do not have a good chance of survival and the ability to pass on their genes

However, in reality, diversity is what gives us a better chance to survive by making us more able to cope with changing conditions.

Imagine if we all had the same exact genes and lived close to each other. We would have all been wiped out long ago by a single kind of virus. Nobody would have a chance to survive if we were all exactly the same. We can see this effect with immediate family or a close-knit group marriage, where the offspring are more prone to diseases and genetic disorder. All complex life forms on Earth reproduce by sharing genes to produce offspring that are a little different from them, and not by splitting their own genes, which would make the offspring almost identical to them.

Imagine if we all had the same religion. You wouldn't know what religion you had because there would be no other religions with which to compare it. You would not be able to relate to right or wrong and your imagination would collapse.

Imagine if we all looked the same, if we were all the same size, shape and color. Imagine if we all ate the same kind of food and spoke the same language, if we all liked the same things and wore the same clothes, if we all thought the same way and understood each other so perfectly that we really did not have to talk. Would this be Heaven or Hell? Actually, this scenario is impossible in our universe, since life is all about interaction with different things, which gives rise to diverse matter structures, diverse ways of thinking, and therefore a better chance to cope and adapt to changing conditions.

Without diversity there wouldn't be anything unique, anything special, any flow of new information. Everything would be the same, time would stop, and everything would collapse.

17. Water

Liquid water is a must for the survival of any life form here on Earth, and it is the common denominator of all life on Earth. It is so difficult to imagine how life forms could exist on different planets without liquid water that we are always looking for signs of liquid water on other planets, and we assume that if we found liquid water, some kind of life similar to life on Earth would follow. Liquid water has special and unique properties. It protects, carries and interacts with organic molecules to form all the many and diverse life forms here on Earth.

In this chapter we (God and I) will explain where water comes from, its special properties, and its connection to our life forms.

Water (H_2O) is made from two atoms of hydrogen circling one heavier atom of oxygen to form a water molecule. Hydrogen is the simplest and the lightest of all atoms. It has only one electron circling its nucleus, which is made from one proton. Hydrogen was the first element that appeared in the universe, and it was created a few thousand years after the Big Bang, when the diverse subatomic particles coalesced. It is the most common type of atom in our universe, and is the source of fuel that burns inside the stars.

A star is created when a huge gas cloud made mostly of hydrogen collapses in on itself because of its own gravity to create a sun with a very strong gravity force. Since this sun occupies far less space with the same amount of matter as the gas cloud from which it was created. The strong force of gravity inside the sun's core creates tremendous pressure that causes collision and fusion of the hydrogen atoms inside the core to create heavier atoms. This formed the diverse array of elements we know of in the universe today, and one of them is oxygen.

So, oxygen atoms are a byproduct of the fusion of hydrogen atoms and are abundant in our universe. Oxygen atoms are easily attracted to other diverse atoms to create many different molecules, the most common of which is the water molecule.

So, water is a common molecule in our universe, but is mostly found either in its frozen/solid or gaseous state. The frozen state is common because most of our universe is very cold, only a few degrees above absolute zero in which water turns to ice. Gas is common because liquid water without atmospheric pressure becomes gas and evaporates, bypassing its liquid state in temperature above the freezing mark. For example, in a comet made from rocks, dust and ice that circles our Sun for billions of years, the cold temperature in the space of our Solar

System keeps the water in the comet in its ice state, and only when the comet gets close to the sun, at about the same distance between Earth and the Sun, the temperature of the comet rises and some of its outer layers of ice evaporate to become gas mixing with dust to form a tail that is many thousands of miles long. Because the comet does not have any atmospheric pressure to speak of, its ice will evaporate, bypassing the liquid state.

The water molecule is a light molecule and is supposed to be in a gaseous state at a certain temperature and atmospheric pressure, but its electric charge is on one side of the molecule since it has only three atoms in it and one of them is the oxygen atom, which is about sixteen times heavier than the hydrogen atom, which means the electric charge is always close to the oxygen atom. So, it binds easily to other water molecules to form a condensed substance, which is either liquid water or ice depending on temperature and air pressure.

So, liquid water is made from a light molecule, and because of that is able to evaporate at any temperature, and by doing so keeps the surface from which it comes at a cooler temperature than the surrounding temperature. This property of liquid water to evaporate at any temperature makes water a necessity for any life form that can function only at a certain temperature range. Water evaporation helps to cool animals and plants alike.

The evaporation of water from the seas creates rain clouds and makes water available almost anywhere on our planet. The rainwater falling on the mountains and the plains carries nutrients along with it as it moves to lower points with the help of Earth's gravity and then evaporates with the help of our Sun's energy in an endless weather cycle that creates more clouds and water vapor in the atmosphere around our planet. The water vapors in the clouds give our atmosphere most of its energy and movement and keep atmospheric temperature as a whole more or less even. Without water vapor and clouds in the air the difference in temperature between day and night and from Earth's poles to the equator would be much greater. Without clouds there would never have been any lightning helping to start the first carbon life form on Earth, which helped to structure the first gene with its energy inside a pool of liquid water.

Another unique property of water is the ability to expand when it freezes. At certain temperatures water molecules create ice crystals with gaps between them. The liquid water becomes ice and expands up to twenty percent more in volume than in its liquid state, depending on the atmospheric pressure and temperature. This ability of water to expand in freezing temperatures makes it lighter than liquid water and it floats on the liquid water or creates a layer of ice on top of the liquid water. This layer of frozen water acts as insulation and prevents the liquid water beneath it from freezing in cold temperatures, and the liquid water beneath the ice is able to sustain life.

Liquid water is a poor carrier of thermal energy. In other words, you need a relatively large amount of energy to change the temperature of liquid water. The fluctuation in temperature around a large body of water is much less than in other places without water. About seventy percent of Earth is covered with water and most of the life forms on Earth contain about the same percentage of water in their bodies, including plants and animals, whose blood contains over ninety percent water.

One of the main reasons life developed on Earth is the abundance of liquid water. Water is a good conductor of electricity and it is friendly to the living cells that communicate and function by exchanging electrons. Indeed, all living cells contain water. Liquid water helps to dissolve and carry nutrients and all kinds of molecules to living cells and carry out waste to keep the living cells clean and healthy.

Liquid water is the delivery medium that promotes the movement of all kinds of substances inside and outside of living cells. Movement, and hence time, means life.

We can make our food last longer and prevent it from spoiling by taking the water out of it (dehydration) or by freezing it, and therefore stopping the movement of organic life.

So, water is a must for life. For your health drink a lot of water. There is nothing like it to promote healthy circulation in your body, to prevent diseases, and to enable long life.

18. Life on Different Planets

Life is the ability of a matter structure to replicate itself more or less in the same way. This means our universe is alive. It was born in the Big Bang about fifteen billion years ago from a different dimension and will die in about fifteen billion years from now, when it will become a different dimension and create many universes that are different from our universe.

The stars in our universe are born and then die in explosions, and new stars are reborn from the leftovers of the explosions of the previous stars. Everything in our universe, including our universe itself, follows the law of time, which says that if there is a beginning, there must be an end, and that end will resemble the beginning.

But when we talk about life as we know it, we are referring to life forms as they are found on Earth, whose instructions are dictated by their genes (or some other molecule that can mimic the same function). Genes are organic molecules because they contain carbon atoms, which are the back bone of those molecules. Carbon atoms in each gene attract and hold other diverse atoms to form a molecular structure that can more or less replicate itself with the help of some kind of energy. Over time and evolution genes attracted other diverse genes to form living organisms that can more or less replicate themselves.

All the diverse life forms on Earth need sufficient amounts of liquid water, the right chemicals in the right quantities, some kind of energy source and plenty of time (billions of years) in order for organic molecules to interact and create these complex life forms through trial and error.

Life appears to be relatively easy on Earth because it has the right combinations and the conditions to develop and sustain life as we know it. We assume that life like ours is possible on other planets. This is true if there is another planet just like Earth with the same exact conditions, such as size, energy source and lifespan.

The odds of having a planet exactly like Earth in our universe are very slim to none, since our universe is limited in size and limited in its number of planets. Although it has a huge number of planets, it is not an infinite number. Our universe is also limited in time. It has a lifespan and a time that is not for eternity, that is not infinite. All this means is that it is certain there are other planets in our universe, some of which will be similar or have close enough conditions and properties to Earth, and can therefore give rise to life forms similar to those on Earth, but not identical.

Our planet Earth is like a bubble of life that encourages and sustains further life production by containing abundant and diverse amounts of chemicals and liquid water at the right temperature. It has the Sun as an energy source that will last for many billions of years. By having a strong magnetic field, Earth protects its life forms from the energies that come from Space in the form of high-energy shortwave photons and high speed particles that would destroy DNA through their interactions. Earth has sufficient gravity to possess a large atmosphere that acts like a protective envelope with enough pressure to keep water in its liquid form.

Now let's go back to our universe to find out where life forms similar to those on Earth might exist. There are over one hundred billion galaxies in our universe of different shapes and sizes, and each one of these galaxies contains anywhere from one billion to one trillion stars. These galaxies also contain a much smaller number of black holes and neutron stars that are usually situated at the center of each galaxy where most of the galaxy mass is concentrated. Galaxies also contain many more white or brown dwarfs and many more suns of different sizes. Life as we know it couldn't exist on those heavenly bodies because their conditions are too extreme, with temperatures of millions of degrees, very strong gravity forces, high-energy shortwave photons, and you can forget about liquid water.

Just as in our own Solar System, around some of these suns are many planets. These planets are made from the same stuff of which each of their suns is made: the hydrogen and dust cloud left over from the Big Bang and previously exploded stars. So the makeup of those planets depends on the makeup of each of those gas and dust clouds. If a cloud contains mostly hydrogen and light atoms and very few heavy atoms, the planets around that sun are gaseous and also not fit for life as we know it. So the newer gas clouds contain a bigger percentage of heavier atoms, which is the stuff Earth is made from. By newer gas clouds, I mean a new generation of gas and dust clouds that have been through multiple cycles of generating suns that have exploded and regenerated new suns and so on. Every time a sun explodes it enriches the gas and dust cloud around it with more heavy atoms.

All this means is that our universe didn't have the right chemistry in the right proportions during the first nine billion years after the Big Bang to create an Earth-like planet. It took about nine billion years after the Big Bang and the creation of our universe before some of the gas and dust clouds had the right chemicals in the right proportions to create solar systems with planets similar to Earth in their properties and compositions that could give rise to some kind of life forms similar to those on Earth.

The main ingredient for life as we know it is plenty of liquid water. Water can stay in its liquid state only in certain temperature ranges (fluctuating about two hundred degrees Fahrenheit) and only under a certain atmospheric pressure, so the distance of any planet to its sun is very crucial. If the planet is too close to the sun its water will get too hot and evaporate, whereas if the planet is too far from the sun its water will turn into ice. The size of the planet plays a big role in its ability to sustain life. If it is too big and heavy, a planet will restrict the movement of life forms because of its strong gravity force. Strong gravity also creates a high atmospheric pressure and powerful winds. However, if a planet is too small it won't be able to hold a sufficient atmospheric pressure and therefore won't be able to hold liquid water, because the water will evaporate.

The other main ingredient for developing complex life forms is plenty of time, so a planet's sun has to be the right size to shine for more than five billion years. A big sun will burn very hot and will last only up to four billion years depending on its size.

The location of the solar system is also important. A solar system in the center of a galaxy will be subject to greater gravitational forces from nearby stars, since the distance between stars in the center of a galaxy is much shorter relative to the distance between stars at the edge of a galaxy. A solar system in the center of a galaxy will also be subject to more shortwave photons and cosmic rays from the explosions of nearby suns and the radiation from black holes and neutron stars. Furthermore, it will be subject to more collisions of stars and planets.

The inner motion and makeup of the solar system is also important. Some solar systems could have planets and other objects spinning in different directions, causing high rates of devastating collisions. Some solar systems could have planets orbiting their suns in an elliptical motion, causing extreme fluctuations in temperature (-300 to 900) and strong gravitational tides.

The solar system itself could be created from a few planets orbiting two suns, or one sun and a white dwarf. Such solar systems change formation and conditions too quickly, making life on their planets highly unlikely because life forms similar to those on Earth need a stable planet without much radiation and with limited collisions of suns, planets, moons and asteroids around them. A long time period (billions of years) of stability is needed to give rise to some kind of life form that can start with the fusing of some molecules through trial and error to produce complex life forms by evolution. A long time of such stability is a rare commodity in our fast moving and violent universe.

So, let's take a look at our own galaxy, the Milky Way, as an example of a place that can accommodate life as we know it. The Milky Way is a medium sized galaxy that it takes about one hundred thousand years for a light ray to cross. It contains at least one medium sized black hole in its center and about one hundred billion stars, most of which are in the center or close to the center of the galaxy. We have to eliminate those stars in the center or close to the center of our galaxy from any hope of accommodating life because of their density and fast moving, violent environments. This eliminates ninety percent of the suns in the Milky Way.

But, there are still about ten billion stars left around the edges of our galaxy. It is not just by chance that our Solar System is at the edge of our galaxy, away from the unstable central areas. Of the ten billion stars left, we have to eliminate a further ninety percent due to improper sun size, solar systems with multiple suns, and solar systems with white dwarfs and giant planets that would eliminate the creation of smaller planets (such as our Earth) that might circle a sun at the right distance. This leaves us with about one billion stars that could have a solar system that resembles our own. Of these, we have to eliminate 99.9 percent of the planets around those suns for not having the right stuff: the chemical makeup resembling that of Earth, the diverse atoms in the right amounts, plenty of liquid water and the right energy levels gained from their suns. This leaves us with one million suns that could have some planets around them resembling Earth more or less in terms of lifespan, conditions and chemical composition.

This large number of potential life accommodating suns suggests that life is possible and probably abundant even in our galaxy alone. Yes indeed. But it may be in its lower forms, like bacteria and viruses. The jump from these organisms to multicellular organisms like algae, grass and worm-like animals is a giant leap forward. This would eliminate 99.99 percent of those planets, leaving us with about one hundred planets throughout our galaxy that could each be a home for higher life forms such as algae, grass, worms and sea animals with no brains.

Now, to make another giant leap forward on some of these planets toward the creation of small animals with small central brains, such as insects and fish, we have to eliminate another 99.99 percent of the planets. This leaves us with 0.01 percent of planets in our galaxy that can develop these kinds of life forms. This means one planet in one hundred average galaxies. There are about one hundred billion galaxies in our universe, but from that number we have to eliminate fifty percent of these galaxies because of their fast moving planets and suns and high rates of collision along with the presence of one or more giant black holes in their centers. This leaves us with fifty billion galaxies. One percent of fifty billion is five hundred million planets throughout our universe that might be able to create and develop a complex life form like an insect or a fish with a small central brain.

Now, to go to a higher life form, to a bigger animal with a bigger brain with some communication skills, which has the ability to live in a community and can plan ahead for the near future, we have to eliminate another 99.99 percent of these available planets, which leaves us with about fifty thousand planets throughout our universe.

The next development is an even higher life form, a complex life form which possess a human sized brain or perhaps one even bigger, which has the power to think in abstract ways, a well-developed ability to communicate and to plan far into the future, and the skill and knowledge to develop and create complex objects and machines. It has the power to be on top of the evolutionary pyramid.

The evolution of nature is a slow process which takes billions of years to develop from a small and simple molecule into a very complex and large structure like the human brain. Therefore, we have to eliminate 99.99 percent of these available planets, which leaves us with about five planets throughout our universe, give or take a few. So, only about five planets throughout our universe are possible homes for intelligent life, and planet Earth is one of them.

By intelligent life form, I mean a complex matter structure that can ask itself questions. Who are we? Where do we come from? What is our destination? What is the meaning of life? Are there any other life forms like us in the universe? Perhaps there are a few intelligent life forms in our universe and perhaps they ask themselves the same questions. Perhaps they know the answers, and the answer to all these questions is that the big brain is a nature-made matter structure that grew in time to become larger and more complex. This gave an advantage of survival to any life form that possessed it by giving that life form the ability to gather information about its surroundings and information about the laws of nature, and to use this information for a better future and a better chance for survival. This in turn meant to better the ability to preserve identity, which is the meaning of life.

Now, who is to say that those intelligent life forms are like us, looking for other intelligent life forms in the universe? Maybe they are happy with their lives and do not want to spend energy looking for other life forms. Maybe there is an intelligent life form out there with less advanced technology than us. They surely didn't have the old Chinese, Egyptians, Greeks, Romans or Einstein (to mention a few in our history) to jumpstart their civilization and science like we did. And they surely didn't have my English teacher.

Maybe there were other intelligent species whose life was cut short by a cosmic holocaust, which is in the future of us all.

Maybe there is a species of intelligent life form that has an advanced technology and has used it to occupy some of the surrounding planets and maybe even the nearby solar system as part of their evolution to preserve identity and to have a better chance for survival. However, according to the laws of nature, it is not possible for them to come to us from a distance of millions and perhaps billions of light-years away. The entire universe has the same laws of nature, and intelligent life forms must know them to become advanced in technology.

So, it is as if we are all alone in the universe, and perhaps this is very true. Perhaps we are the first ones. An older universe, perhaps millions or billions of years from now, might be more accommodating to other intelligent life forms.

Even planet Earth, with its perfect conditions to start a new life form, still took about one billion years to create the first gene, after which diverse genes grouped together to form the simple life forms of bacteria and viruses . It took Earth another four billion years to create the human being with the largest brain on the planet, who can think in abstract ways and ask the questions: who are we; where do we come from; what is our purpose?

It is clear that nature works very slowly. It does things by chance, by trial and error, and it hits a lot of dead ends. If Earth's conditions had remained the same from the beginning, it could have given rise to a single successful species that would stay about the same through the passing of time, without becoming more complex, advanced or diverse.

However, Earth's conditions did change with time due to volcanic eruptions, climate changes and extreme events. One such event occurred sixty-five million years ago when a large asteroid struck Earth, destroying ninety percent of all its species and almost all the dinosaurs. These changes drove the direction of evolution of life on Earth. The extinction of the dinosaurs allowed mammals to flourish and gave rise to Homo sapiens, from which human beings later on developed. On the other hand, if this asteroid had been larger, it could have destroyed all the life forms on Earth, and new life forms would have had to restart again from scratch, maybe in a different form, or simply, Earth could have been left lifeless.

It is like a game of Russian Roulette and other planets face the same odds. Chance events drive the evolution of species, and one of those species that survives can maybe later in time give rise to intelligent life forms. Some species might be eliminated completely by viruses, bacteria or successful predators that target them specifically. Also, as a life form becomes more complex, it becomes more fragile.

For these reasons, I give the odds of ten thousand to one against the existence of a higher life form in any stage of its evolution. New life forms always start with small numbers (many times from one) and simple forms to become more numerous, diverse and complex with time, during which their conditions have to change constantly but not too drastically.

To summarize, life forms can and do exist in extreme environments without liquid water, but they can only evolve into existence with the right conditions and plenty of liquid water.

19. Life on Earth

Earth is one of a number of planets that circle the sun in our Solar System. Our Sun is one of billions of suns that have a solar system in our galaxy, the Milky Way, which is one of billions of galaxies in our universe. Nature throws very big numbers at us to beat the odds of finding a planet with the perfect conditions to start a new life form and to give this life form enough time for evolution to drive diversity and complexity.

Earth is only one of millions of trillions of planets in our universe and was born about five billion years ago as part of our Solar System, which came from a gas and dust cloud with lots of rock debris that was left over from collisions of planets and dying suns and the explosion of a large sun somewhere close to the center of our galaxy. This large sun exploded as a supernova when its fuel (the hydrogen and helium atoms in its core) was exhausted. The explosion created a hot gas cloud containing only the atoms that were made inside of the dying star. Anything bigger than an atom or a stripped atom couldn't stand the heat of the explosion and the initial acceleration speed of about fifty percent of the speed of light. These atoms traveled toward the edge of our galaxy via shockwaves created by the exploding star, carrying with them any leftover gases, dust clouds and rock debris in their path.

The journey of these atoms took many millions of years as the gas cloud pushed other gas clouds and Space debris toward the edge of our galaxy. The atoms in the cloud attracted other diverse atoms through chemical reactions whenever they had contact, with the help of shockwaves, collisions and electrical attractions. This process formed new molecules, creating mostly water, iron oxide, silicon and carbon. Later on, these carbon atoms fused with other atoms to create bigger molecules of amino acid, which is one of the most important building blocks of life on Earth. As this gas cloud made its way toward the edge of our galaxy, in about half a billion years it cooled down, slowed and condensed.

The big rocks inside the center of the gas and dust cloud attracted smaller bits of rock, gas and dust around them with the force of gravity. As this condensed gas cloud became bigger and heavier it gained a stronger gravity force, allowing it to wipe clean the area surrounding it for many millions of miles. It became big and heavy, with a tremendous amount of pressure inside of it. This pressure became great enough to fuse the hydrogen atoms inside of its core because of its own tremendous gravity force. It became a sun that now produced heat and energy in the form of photons to heat and light up our new Solar System for many billions of miles.

In the meantime, there were also a lot of small and large rocks circling the new Sun many billions of miles away. These rocks also got bigger by attracting other rocks, dust and gas around them through collision and gravity force, and they became the planets of the Solar System, while the smaller rocks became their moons if they were close enough to each other. It took another half a billion years for our Solar System to look more or less like our present Solar System. It took this long for collision and gravity force to clean most of the Space debris in the inner circles of our Solar System. The outer circles where the Oort cloud is located, and made mostly of small rocks, were much denser and wider than today's Oort cloud.

So, Earth was born as a part of our Solar System in about half a billion years from a gas and dust cloud that formed somewhere near the center of our galaxy through endless collisions, motion and gravity force. Earth, the third rock from the Sun, has a perfect distance from the Sun to enable it to have the right temperature for its water to be mostly in a liquid state. Earth also has the right size to create a gravity force to hold atmospheric pressure that keeps water in a liquid state at the right temperature. Earth has a perfect size that creates enough gravity force to possess a large atmosphere and a strong magnetic field to protect its life forms from cosmic rays, high speed particles, dust and small rocks falling from Space. Earth has enough gravity force to keep a relatively large moon nearby that adds stability through joint gravity against other large bodies in the Solar System and gives extra protection against collision with Space debris. Yet Earth is not too large and heavy to create a gravity force so strong that it would restrict the movement of life forms or possess a high atmospheric pressure with very strong winds. Earth has the right stuff. It has all the diverse atoms that were made inside the previously exploded sun: plenty of iron, carbon, oxygen, nitrogen and hydrogen.

Not all planets spin, and some spin too fast or too slow. Earth's spin gives us day and night with enough time for adjustment between the two different conditions of light and darkness. The Earth tilts as it circles the Sun, creating four seasons whose changing weather conditions help diversity of life forms. Changing conditions and diversity accelerate life's evolution. Earth is a part of a stable solar system at the edge of our galaxy, far away from black holes, neutron stars and exploding suns with their shockwaves and killer cosmic rays. The nearest solar system is four light-years away from ours, which is pretty far.

Our Sun is a medium to small sun that will last over nine billion years from the time it was created: plenty of time for nature to start life on a planet with perfect conditions, through trial and error. And it is plenty of time for this new life form to grow and become complex and diverse through evolution, without God's intervention.

Earth was a hot place the first half of the billion years of its creation because of the endless collisions with small and large rocks, asteroids and comets, during which time it grew in size to the size it is today. The young Earth's crust was thinner than it is today; it had a lot of movement and lava flowing in many places. Earth had a hot and thick atmosphere that was made mostly of water vapor, carbon dioxide, nitrogen and methane. These greenhouse gases kept the atmosphere hot and stormy with a lot of rain and lightning. But the conditions on Earth were ripe to start life in its simplest form, as bacteria and viruses that developed from the right conditions and the abundance of chemicals and liquid water.

Organic life is after all an extension of the chemical reactions that happen everywhere in nature all the time. Diverse atoms attract other atoms by electric charge to form simple molecules, and molecules attract other molecules to become larger and more complex. All life forms on Earth are organic because of the presence of carbon atoms in their molecules. The carbon atoms are the backbone of all organic molecules because of the strong attraction between them that allows them to form a long chain of molecules. (Diamonds, made entirely from carbon atoms, are the hardest substance in the universe.) These carbon atoms can attach to other atoms of nitrogen, hydrogen and oxygen to form complex molecules including amino acids and nucleotides, which are the key to the biochemistry of life.

These molecules can be created in Space and on Earth, but on Earth the help of liquid water provides the medium to bring or carry them to an extra source of chemicals, and with the help of some kind of energy source which acts as a catalyst, like hot lava or lightning, these molecules can fuse together and create many different molecules with the further help of time (in millions of years). One of these molecules was the first gene, an organic molecule that can replicate itself the same way under the right conditions.

This first gene ultimately evolved into DNA, which is seen in all life forms on Earth, from the small and simple viruses and bacteria to the much larger and complex plants, animals and us, humans.

20. The Gene

The first successful gene appeared on Earth about four billion years ago. A gene is an organic molecule that can replicate itself and also produce a substance, a protein that can protect the structure of the gene from outside elements, and by doing so preserve the molecular structure of the gene.

The first gene copied itself under the right conditions and later on in time attached to its own mutated genes to form DNA in the living bodies of all life forms on Earth. At the time life began, Earth was already about half a billion years old and existed as an integrated part of our new Solar System at the edge of our galaxy.

When the first gene was created Earth was a barren, stony and dusty place, and the bombardment of objects from Space had slowed down. Earth's crust was thinner than it is today, with lots of movement and lava flowing at many places, and with many hot springs. Earth had a large sea and hot stormy weather with a lot of rain that created streams and rivers flowing to the low points by way of Earth's gravity to become ponds and lakes which contained chemicals that reacted to create many new and different molecules. After many millions of years of creating new molecules through trial and error, with the right conditions, right chemicals, plenty of liquid water and plenty of energy coming from the Sun and the fusing power of Earth's gravity, lightning hitting the water and the hot lava flowing into water, a new molecule appeared on Earth. Although the exact structure of this new molecule is not known, it had a backbone of carbon atoms, but what was unique about this new molecule and set it apart from other molecules is the fact that it could replicate itself under the right conditions in exactly the same way, along with a second unique property of this molecule which is its ability to produce or use a substance that acts like a shield to protect it from outside elements and only let in desirable atoms or small molecules through chemical reactions.

Without these two unique properties together, the gene wouldn't be a gene; it would be just another molecule that comes and goes without the ability to preserve itself for long periods of time by replicating. And that is exactly what happened in the early days of the creation of life on Earth. There were many new combinations of molecular structures (including those which had carbon atoms in them) that couldn't preserve themselves and therefore hit a dead end. The gene is a long chain molecule with a backbone of carbon atoms which have a strong attraction among themselves that creates a molecule with a strong bond, a molecule that includes many atoms of oxygen, hydrogen and nitrogen.

So, the gene is basically a molecular structure, and as any structure, it is also an information unit, just like a word that is made from letters. The molecular structure of the gene contains the necessary information to create another molecule of protein. The same gene interacting with the same chemicals will always create the same protein, just like the same word, which is structured by the position of its letters, will always put out the same information.

The English language contains twenty-six letters. Each letter is different from the others and contains the information to make a different sound. In general, you need two or more letters to make a word that means something to somebody. This is just like atoms and molecules. You need at least two atoms to make a molecule, which is also an information unit.

So, the relative position of each letter in a word is very important. As a matter of fact, it is everything. If the relative position of any letter in a word is messed up, the word will have a different meaning or could be rubbish with no meaning. In the same way, the relative location of the atoms in a molecule is important, and in the molecular structure of the gene they will copy themselves in the same way to have the same meaning.

Sometimes different genes have the same kind and the same number of atoms in their structure, but they position them differently and therefore give different information. Just like the words "time" and "mite" have the same letters and the same number of letters, but because the letters are positioned in different relative spaces within the word, the two words put out totally different information.

In the same way, every atom and its location in the gene molecule has a meaning, a bit of information. But unlike the English letters which are positioned in two dimensions of space, the gene molecule positions its atoms in three dimensions, and this extra dimension of space gives the gene molecule more than twice the amount of information relative to two-dimensional space, so it can pass on more information to the next generation or create a protein molecule with amazing details. The gene molecule contains about sixty bits of information, and unlike the English language that has only twenty-six letters, all this information of the gene is packed in a molecule that is much smaller than a pinhead. With only twenty-six letters in the English language we write all the books, newspapers and encyclopedias that exist, and that's a lot of information coming out of just twenty-six letters, created just by the maneuvering of letters to different relative positions in order to spell words. So with sixty letters, the information can be more precise. So, this little gene can pass on a huge amount of information.

For example, the group of genes that are responsible for hair can grow hair exactly as they are programmed to, with amazing details of size, color and shape all the way down to the exact molecular structure of hair. Genes store a lot of information because they always build a molecular structure in three dimensions, and you need a lot more information to build a structure in three dimensions than in just two dimensions.

Human DNA contains over two billion units or bases, which encode about thirty thousand genes. The total size of a DNA molecule is smaller than a pinhead, but it contains enough information to build the complex structure of a human being from the inside out. The miracle of creating a new body with genes is actually simple, but since it involves a huge number of genes and a huge number of interactions and communications between those genes through chemical reactions, and all this is done on a very small molecular scale that we cannot see, for us it looks like a miracle.

The creation of a new body goes according to nature's laws and it works as follows: we know that the gene creates a substance, a protein, and that different genes create different proteins, which are the building blocks of the living body. The different genes work in harmony and by sequence, with synchronization, timing, and therefore order, which creates intelligence as a reflection of the order, to create a new body. Some genes will create their protein only after some other genes have created their protein, as they get triggered by the presence of the new proteins to create their own different protein. On the other hand, some genes will stop the creation of their protein in the presence of certain proteins.

For example, some genes in the subconscious part of the brain will produce a substance that will create the feeling of sexual desire only in the presence of certain hormones in the body. These hormones are a product of other groups of genes that were activated when the body reached a certain level of maturity: a chain reaction of a chemical reaction.

So, the first single successful gene molecule appeared on Earth about half a billion years after Earth was created. It was developed in a big body of murky water as a result of chemical reactions involving organic molecules driven by the energy from lightning striking water or from hot lava flowing into water to create this new molecule.

The first gene was a small, simple molecule that had very strong attraction between its atoms and had the water's mercy to either carry it or to bring other atoms and molecules to it so that it could constantly create a shield, a substance made out of protein, which is also an organic molecule that protected it against ultraviolet light and the harmful chemicals in water. All this was done without either God's intervention (only with God's blessing) or an intelligent design. Such a shield is created through a chemical reaction that is influenced by the molecular structure of the gene and is subject to the laws of nature that put order and therefore intelligence into every structure in our universe. The design of the shield happens to protect the gene structure and only let in the right atoms so that the gene can copy itself and keep producing new shields. If the shield couldn't do those tasks, we wouldn't be here to ask questions.

Genes have a lifespan, and just like any other matter structure in our universe, the lifespan depends on outside conditions, and in the case of the gene, also on the number of times it can produce its shield. This shield can only be produced a certain number of times before the gene disintegrates and the information of how to build the shield vanishes. But, before the gene structure disintegrates, it is able to copy itself a few times by allowing the same kind and the same number of atoms of which the gene molecule is made through its shield, and positioning them in its exact relative position like a mirror image, a reflection of its own structure. In other words, it allows passage to the same kind of atoms in the same number and position as that gene's molecular structure, and by doing so creates another gene that can create its own shield. That shield's molecular structure will be the same as the previous gene shield, and it will disconnect itself from the mother gene to become independent.

So, in the timespan of a few days, this first gene, the mother of all genes, successfully copied itself a few times before it disintegrated. The offspring genes then copied themselves a few times and they multiplied to a few million genes in just a few years, although many million more genes didn't make it and died without propagation. The genes that survived were carried by water to different destinations to face different conditions of temperature, exposure to ultraviolet light and the availability of chemicals in the water that carried them.

Some of those single genes didn't copy themselves in their own mirror image and produced a mutation of their own gene structure due to their interactions with the new conditions of their surroundings as they went through different locations and times. The newly created gene, which had a different molecular structure from the mother gene, didn't split apart from the mother gene, but instead attached itself to the mother gene and created the first living cell that had more than one single gene in it.

So, in a timespan of just a few million years there were billions and billions of those single cells with increasing numbers of mutations (alterations) that created diverse cell types. Some of these cells created different shields around themselves (made out of protein) to protect them from their own environments and bring in diverse atoms so they could reproduce themselves.

Adding a new gene to the cell could cause two things to happen. In one scenario, the newly added gene creates its own shield, its own protein, that interferes with the functioning of the mother gene's protein, and both genes hit a dead end wherein they are unable to survive and reproduce. In the second scenario, the new gene creates a shield that works hand in hand with the mother gene's shield, enabling the new cell to reproduce and survive in more diverse conditions, or to reproduce better than the mother gene alone.

By the way, this trend of adding new genes to living cells continues to this day. Sometimes it helps the body to survive new conditions, but more times it leads the body to a dead end. By grouping together, genes had a better chance of survival since they could withstand more diverse conditions and use more diverse atoms and molecules to reproduce and build new and improved shields for protection.

So, the second stage of life forms on Earth resembled the viruses, with their small size and the small number of genes in their DNA. And as time went forward in a span of many millions of years a new life form appeared on Earth, the third stage of life forms: mold and bacteria, resulting from the grouping of many genes and viruses under Earth's right conditions.

Again, all this was done through a huge number of chemical reactions and without intelligent design (maybe by so-called "stupid design"), only by trial and error and by being in the right place at the right time. Some of these new organisms made it to become larger and more complex life forms of bacteria and mold, and many more groups didn't make it, hitting dead ends and disappearing without a trace. The DNA of these new life forms contained more than a hundred different genes, and their number swelled to many millions of diverse species of bacteria and mold consuming a wide variety of available molecules on Earth for the purposes of their existence and reproduction.

Some of these new life forms consumed sulfur from the water, carbon dioxide and methane gas that were abundant in the atmosphere in the early days of Earth, and some started to occupy the land near the water from which they emerged. Some of the mold became algae and started to use the Sun's energy to put together water molecules with carbon dioxide. Their byproducts were carbohydrates and oxygen created through the process of photosynthesis. With the passing of time, in billions of years some of the algae became green grass and plants that competed with each other for sunlight by becoming bigger and taller and occupying more space on land wherever there was enough water, dispersing their DNA by floodwater and wind to occupy new territories and consuming methane and carbon dioxide gas from the atmosphere to change the atmospheric balance of gases to mostly nitrogen and about twenty percent oxygen.

Oxygen produced ozone, which is a light gas, and occupied the upper level of the atmosphere. It has the unique property of reflecting ultraviolet light coming from the Sun back to Space, and by doing so protects some life forms on Earth from high-energy ultraviolet light that can damage the structure of their DNA once they are out of the water. These changing conditions on Earth didn't go unnoticed, and many life forms took advantage by adapting to the new conditions. They became more complex and grew bigger to contain many thousands of genes, while many ancestral genes went dormant. In other words, some of the genes didn't produce their protein under normal conditions; those dormant genes only produce protein when the conditions call for it.

So, about three billion years after the first gene appeared on Earth, or about one billion years ago, there were many millions of diverse life forms on Earth, many with over ten thousand genes in their DNA. Some of the green algae evolved into plants and trees, changing Earth's climate to become cooler and giving rise to a variety of life forms such as bacteria and other multicellular organisms that fused together through symbiosis to become bigger, diverse and more complex, such as insects and worms. Some of the new life forms left the water and went inland, consuming carbohydrates (generated by planets) and oxygen from the air for their energy. About half a billion years ago there were over a billion different life forms occupying almost all of Earth's waterways and land surfaces wherever liquid water was available.

The explosion in the diversity and number of new life forms on Earth created competition for space and energy, and many life forms adapted to these new conditions by becoming more diverse, by creating new genes and possessing more genes in their DNA to become bigger, faster or with the ability to reproduce more quickly and in larger numbers. Some adapted by consuming other life forms, which in turn gave rise to all kinds of body sensors and body parts for protection or to attack other life forms.

There were many setbacks in the creation of life forms on Earth, such as huge volcanic eruptions or the collision of large asteroids with Earth. Some of those events wiped out a large percentage of the bigger and more complex life forms on Earth just to give rise to different life forms that could adapt to the new conditions and use the new opportunity of more space and less competition to fill the empty gaps by forming new and more complex life forms with over ten thousand genes in their DNA.

Some of these new life forms developed brain cells that enabled them to be aware of their surroundings by storing and comparing information from the past to foresee and calculate future events. This gave them a better chance of survival in the competition with other life forms for space and energy. The competition for space and energy along with constantly (but not too drastically) changing conditions, plenty of liquid water, the right chemical components and plenty of time (millions and billions of years) enabled the evolution of life on Earth.

The competition for energy and space among the new life forms, along with their ability to adapt to changing conditions, is the main force behind evolution. The desire to survive is very strong, because if you do not survive or are unable to pass on your genes, you lose your identity and are out of the game of life. You can survive by being in the right place at the right time, or by being strong, with the ability to adapt to different conditions, to pass on your genes and to take care of your offspring. By doing so, you preserve your identity longer.

The evolution of life forms on Earth continues to this day and will continue into the future for as long as conditions permit. The first gene is imbedded in the DNA of all life forms, along with some of its mutated genes. There are over thirty thousand mutated genes and their number will grow in the future.

The combinations of some of these different genes along with their different numbers and sequences in DNA make each species different from other species, and each individual unique, different from other individuals within the same species. Just like God said, "You can only be that for which you were designed." A dog will look and behave like a dog because of the combination of his genes. His DNA builds him this way. He won't be able to fly even if he eats bird seeds, because the blueprint of his DNA built his body as a dog and his DNA does not have the information to build wings. Genes design your body to eat certain foods, and if you eat different foods you will suffer the consequences. This is just like a lion who is designed by his DNA to eat only meat and will starve on a grass diet. Even though he shares some of his genes with a cow, his body design is influenced by a different combination of genes to build his body the way it is. So your body is designed by its genes, but the genes can also be influenced by the body's needs.

For example, let's say that most of the time you need to use over ninety-nine percent of your legs' capacity. You do a lot of walking, running and climbing, so the groups of genes that created the bones, blood vessels and muscles in the legs work harder and faster to replace the old and overused cells in your legs more often. Because the lifespan of these cells is shortened due to more interactions, they work harder, and the genes have to produce more of these cells faster. These overworked genes do not go unnoticed. They change their structures by becoming bigger and create more copies of themselves with the possibility of mutating genes over time. So when you pass your genes to the next generation, the groups of genes that are responsible for the growth of your legs will produce stronger legs in your offspring. They won't be much different than the legs you were born with, and there will hardly be a noticeable difference in just one generation. But if all the following generations use over ninety-nine percent of the capacity of their legs, after about two thousand generations there will be a big difference and the people of that generation will have much bigger and stronger legs.

The same is true about the size and capacity of the brain. The more we use it to its full capacity, the bigger and more complex it becomes. The human brain today is double the size of the human brain of two thousand generations ago, or about fifty thousand years ago. So it is a good idea to pass on your genes when you get older because your genes will contain more information, although you might have more complications involving fertilization or birth defects.

Bacteria can adapt to new conditions by changing some of their gene structures within a few months. Since they can multiply (reproduce) in a few minutes, they can have two thousand new generations within a few months.

This process of changing some of a gene's structure in the DNA of living bodies to accommodate the body's needs or changing conditions also works in reverse. The body parts that are either not used or are only used to less than ten percent of their capacity will get smaller, weaker or disappear altogether after about two thousand generations. And the group of genes that are responsible for the development of those body parts will become smaller or will become dormant, and will only be reactivated when changing conditions call for them.

So it is true that if you do not use it, you lose it. It is all about saving energy, because energy is limited and you want to survive by getting rid of things that you do not need or use.

Human DNA contains about thirty thousand genes, some of which are dormant, meaning they seem to not be playing any part in the body's functions. They are left over from an early era, before the human being was fully developed, and played a role in the bodies of a different species that later on in time developed into the human species through group splitting and the mutation of some of their genes. So, some of the old genes become dormant because they were no longer needed, but sometimes some of those dormant genes flare up, activated by unusual conditions, and do what they used to do. They can change the behavior of the human to be more like that of the animal from which it came, or they can change the human shape by growing extra hair or a tail. It happens rarely, but it does happen.

DNA is like an open book of history. It can tell you who are you, where you came from, who your relatives are, and can trace your roots all the way down to the single gene that is the mother of all genes. The gene's molecular structure is, after all, an information unit of past conditions and information about past life forms, and the duplication of the gene brings the past into the present time. The DNA of all life forms on Earth contains the first gene and some of its mutated genes that were developed later on in time. The number of those different genes and their relative positions in DNA constitutes the blueprint of how to develop the body of different species and different individuals within the same species.

Some genes can mutate by way of exposure to molecules or atoms with strong chemical attractions that change the molecular structure of the gene. A gene can also become mutated by exposure to high-energy particles or shortwave photons that can knock out some of its atoms and thereby change its molecular structure. The mutated gene can produce a new substance made out of protein. It could be a tumor or a cancer cell, which in this case could kill the host and lead to a dead end. New mutations in genes could create a new body part, and by improving and using this new body part over many generations they could create a new species that is different than the species from which it evolved in shape, function or behavior. Sometimes, DNA can incorporate outside genes, such as virus genes, to its structure.

It took about four billion years for the evolution of life through the evolution of genes to create the human species, which has the most powerful brain in the animal kingdom. This brain put the human species on top of the food chain. Thanks to all the necessary events and long lasting and stable Solar System, nature had enough time to work by trial and error so that life on Earth could form through chemical chain reactions and nature's design. This took many millions of years. Nature always starts small and simple, becoming larger and more complex under the right conditions.

Through knowledge and brainpower we can now redesign genes to serve our purposes of living longer, warding off disease, and growing food or even body parts, all through genetic modification. We can now do all that much faster than nature, which can take many thousands of years to create a profound change through natural evolution.

Furthermore, we are on the verge of creating a new life form through advanced technologies and Nano science in machines with artificial intelligence, self-sustaining machines with the ability to repair themselves and learn by themselves. And once those machines are able to reproduce by themselves, we can create new non-organic life forms that can have a lifespan of thousands of years, that can withstand much more diverse and extreme conditions than organic life forms, and that can through their evolution occupy much more space in our universe, and by doing so disperse the message throughout the universe that a complex and intelligent life form was created on Earth against incredible odds.

The questions is: is anybody listening; does anybody care; is anybody there?

21. The Brain

As life progressed on Earth there was competition for limited space and energy between the huge number of different species and also among individual members of the same species.

In such a competitive environment only the strongest survive. These different species had to adapt to changing weather conditions, the variable availability of food and sometimes to the arrival of a new predator on the block. Some life forms couldn't adjust to changing conditions and disappeared altogether. Some were consumed by other life forms, and also disappeared.

The evolution of increasingly bigger and more complex life forms (and the genes that created them) gave rise to new cells types: the sensors that allow a life form to detect the senses of touch, smell, sound, vision and electric pulse. Now, all the cells in the living body are sensitive, more or less, to these senses, but only the sensors specialize in detecting those senses.

Let's take the sense of vision as an example. Every cell in the body can sense photons of certain wavelengths as heat, but only the eyes, which are vision sensors that have photoelectric cells that turn photons into the pulse of electrons, can tell the difference between wide wavelengths in the spectrum of light to detect colors and also to pinpoint the direction from which the light comes.

All these sensory cells need a connection to the body through the peripheral nervous system to a central nervous system that controls the whole body's movement. And, for this reason, the simple brain was created. This first brain was born after about three billion years of the evolution of life on Earth to give some kind of advantage to the life form that possess it for survival, so that it could find food or mates or escape predators.

The early brain was very small: about the size of a pinhead. It didn't have any memory cells in it except for genetic instructions on how to grow, function and behave. It operated in simple life forms. But as time went forward, so did the evolution of the brain. Competition for survival became stiffer. Life forms became bigger in size and more complex. They used more and better sensors. And all this put pressure on their brains to grow in size and to perform more complex operations, which in turn gave rise to memory cells inside the brain. That was a giant leap forward.

Memory cells are very important. They let the brain recognize objects by comparing memories to each object, and also enable the brain to see/predict the future based on experience, by having the impressions of the past. To analyze and pick the right memory, the brain must also have analytical power.

Memory cells get information from the sensors and make molecules out of them. Electrons that come from the sensors via the nervous system reshape available molecules inside the memory cell in certain positions, that is to say, the atoms in the molecules reposition themselves in a certain order to create information molecules. Whenever the brain needs that information, it sends electrons via neurons to a memory cell to get the exact impression (copy) of that certain memory in the form of an electron's pulse, which also passes through the neurons. It brings that impression back to the analytical department of the brain for comparison, and that memory or event comes back to life in the present time. The analytical department can compare these memories from the past to the present information it gets from its sensors and order the body's movements accordingly.

Memory is made from atoms put together to form a molecule, just like the gene that stores information by way of the relative position of the atoms in that molecule's structure, and passes it on as impressions of that molecule to make its own copy. But unlike genes, which can take a few minutes to duplicate themselves and are permanent (they do not disintegrate), memory molecules can disintegrate. A memory molecule makes its impression in a split second, because the raw material and the carbohydrates (its energy) are close by and can be triggered by electron pulses from the analytical department in the brain which is not too far from the molecule (only a few inches in the human brain and even less in the smaller brains of other animals).

To understand memory you have to understand what time is all about, because when you use memory it is as if you travel back in time. You can hear, smell and see things that happened in the past, all because of the certain shape of a molecule or the way the atoms order and position themselves in that molecule. That's exactly what time is all about: the relative position of matter structures, be they atoms, molecules or planets. If they are in the same positions relative to each other as they were before, the time or the event will come back. The memory cell holds the molecule it receives from the department of the brain that makes a molecule out of the information it receives from the sensors, and the brain stores the molecule in one of the many memory cells. The whole brain works on the same principle, creating and disintegrating molecules or fusing atoms and molecules with electron pulses. All of this happens in a split second and in many places in the brain at the same time.

Let's take a look at the vision sensor again. When the eye sees an object, or to put it more correctly, when the photons coming out of that object hit the eye, the eye turns the photons into electron pulses. The electron pulses travel via the optic nerve to the analytical department of the brain, where the electron pulses reshape available molecules. And each new molecule or chain of new molecules that was created from the latest electrons coming from the eye makes you think, makes you see what the eye sees. It is like the brain reads the spelling of a word. And all this is done through chemical reactions, all according to the laws and possibilities of our universe. The molecule holds the information just like a word, by way of the relative position of the letters in the word, or by way of the relative positions of the atoms in the molecule. So it is possible that your brain can see different things than what your eyes really see. But since the brain is really you, it is really the judge. You think that what you see in your brain is the real reality. Your brain does not see the outside universe; rather, it creates the outside universe by way of chemical reactions inside of itself. If the brain is not synchronized with external reality, in other words, if the brain is not able to reflect outside of the reality inside of itself, this will lead to a dead end.

Bad genes, accidental injury, disease, plain old age, drugs or alcohol can mess up the process of receiving electrons from the eye, or the process of building molecules in the brain to replace that vision. This is true for all the information coming from all the sensors in the body.

Most of those new molecules that were created to reflect the information coming from all the sensors in the body disintegrate after a few seconds because the events weren't necessarily important and therefore the brain won't have any memories of them. The remains of those information molecules will be recycled in the never ending process of creating new molecules and disintegrating them, as long as the brain is alive.

The main force behind the behaviors of all life forms are genes. Genes and the way they are laid out in DNA build the body the way it is. A tree is a tree because of its genes, and life forms that have a brain follow certain behaviors because their genes produce molecules in their brains that make those life forms behave the way they do. A tiger behaves like a tiger even if he dresses like a lamb, because his behavior is dictated by his genes.

A newborn baby will start looking for his mother's milk and later on in life will engage in sexual behavior without any experience or memory of those acts. He will know what to do, his behavior or actions will come as an urge or a push within him to act the way he acts. This all comes from the genes that build some molecules in his brain in a certain order, and these genes act as his memory cells. They pass the knowledge of his evolution from the past to the present as collective memory.

So, the brain reflects all the knowledge and experiences of the many generations and many different species up to this point in time, including the daily experiences that are embedded in the memory cells of each individual making each of us unique (one of a kind).

22. The Human Brain

After about four billion years of the evolution of life on Earth, the human brain developed as a result of the continuous evolution of the primitive brain. The primitive brain developed about three billion years after the first gene appeared on Earth. The primitive brain grew in size from that of a pinhead a billion years ago to that of a cantaloupe in human beings today.

Nature works through a slow process of trial and error and circumstance, by necessity and with a lot of dead ends. Sometimes the process of evolution regressed or was on hold for a long time, and didn't move forward until an event or change happened. One such event is an ice age, which is a slow process that can take thousands of years of very cold weather creeping from the Earth's poles inward and forcing life forms either to migrate or to adapt to the colder weather. In these events some life forms survived and some didn't, which gave rise to new life forms. Every once in a while a new predator appeared and wiped out some species. The predator could be a fierce meat-eating animal or a virus or bacteria. Earth was impacted constantly by asteroids that were circling around our Solar System, and every few tens of millions of years a giant asteroid hit Earth and changed its climate in a matter of a few days. This would be enough to eliminate most life forms because they didn't have enough time to adjust to the new conditions. However, this in turn gave rise to an opportunity for other life forms to flourish.

The last big collision happened about sixty-five million years ago and destroyed about eighty percent of all life forms and over ninety-nine percent of all the dinosaurs. The dinosaurs' families ruled the Earth for hundreds of millions of years. They were a successful species with its own diversity that relied mostly on size and shear power for survival. Most of them had small brains that didn't get much stimulation and therefore didn't grow much in size and function for hundreds of millions of years. Their almost total disappearance gave opportunities to other branches of life to flourish and become more diverse. Among these were the mammals.

These small, agile and rapidly adapting mammals grew in size and diversity, and so did their brains. Since they mostly lived in groups to have a better chance of survival, they needed to improve communication skills among themselves, be it through smell, voice or body language. All this led to pressure on their sensory systems and caused their brains to grow, so they would have more capacity to communicate and understand each other.

The mammal family branched out many times, and one of these branches were the apes. Apes always lived in groups, and when a group got too big it split, and sections of it went to other territories to find food or to escape the pressure of a large number of the same species. All these groups had to have a leader to protect and lead them. First, was the family as a group, where the mother was the leader of her offspring. She had to provide food, protection and relate some survival skills to them. She lived in a much larger group that was usually led by a large, strong, mature and smart male who protected them against rival groups. He protected their territories, found food for his group and chased other males that tried to take the lead. All this put pressure on his brain to remember things and to outsmart the others. His reward was more food and the opportunity to mate with the females in his group to pass on his genes to more offspring than other males in the group, and by doing so improve the gene pool of that group. This included the genes responsible for brain growth and function.

This process continued for many millions of years, and those groups of apes scattered almost all over the continent of Africa, the birthplace of man. Some of these groups started to look and behave differently than other groups, depending on their food supply, location and conditions. Among the most successful groups of the ape families were the ones that consumed a wide range of food, be it fruit, leaves, seeds, roots, insects, dead animals or dead fish. They were opportunists, eating anything they could catch, such as bird's eggs and live young or injured animals. This high-protein diet and constant food supply, in addition to constant use of their brains for food gathering and communication with their large group, put pressure on their brains to grow in size and function capability.

A few million years later, Homo erectus appeared on Earth as a new evolutionary branch of the successful apes. These humanoids no longer used the trees and were walking on their hind legs. They were standing up to see where they were going and to watch for predators. After many thousands of years their legs became stronger and they were walking upright on two legs. By doing so, they had their hands free to use tools made from stones, bones and wood, in order to ward off predators and also to hunt live prey. They already had the biggest and most complex brain in the animal kingdom, about a quarter of the size of a human brain today.

About a million years after the first appearance of humanoids as a branch of the apes, they had several of their own branches. Many different evolutionary groups split and looked for different territories. They already had a different look and size from other groups depending on their location, food supply and gene pool. Some had more sweat glands than others and some had less body hair than others, so their skin colors adjusted to block or absorb sunlight.

About a million years ago some of these different groups of humans started leaving Africa, heading north and east along the sea shores of the Mediterranean and the Red Sea and through the Middle East to conquer the new territories of Asia and Europe. Some groups stayed in Africa, and later on some of them migrated more or less along the same routes. The conquering of these new territories was a long process that took many thousands of years. They walked maybe fifty to one hundred miles to a location that could provide them with food and water, and they could stay in that location for generations until the place no longer provided enough food for all their offspring. At that point, they either all had to leave or the group split. Some left and some stayed in the same location for more generations.

This process repeated itself in many places. These splitting groups could be as small as one family, or a few males with a few females, or as large as hundreds of them, including their children. Some groups of humans didn't make it. Some died of starvation or disease, or were killed by rival groups, and their gene pool died with them. Life was very harsh in those days.

As time progressed, so did the human brain. About fifty thousand years ago it grew in size to about half the size of today's human brain. The humanoid's vocabulary improved from a few words to hundreds of words, and they could transmit information through voice more precisely. They started to draw on rocks, illustrating experiences which were usually their encounters with animals. They started to domesticate animals by taking young animals away from their mothers, killing or chasing away the mothers and bringing the young animals back to their own communities to play with and stimulate the brains of the children as live toys, and later on to be consumed if food was needed. They started agriculture by weeding out unwanted vegetation while letting desired vegetation grow freely.

Humans got a big boost in their evolution when they learned to harness fire. First they learned how to keep it burning, and later on they learned how to start it. Fire gave them a big advantage over animals, as they could ward off predators and clear a new habitat by burning it. With the help of fire they could build better weapons and tools made from bronze and iron. With the help of fire they could cook food that was otherwise too hard to chew or digest, and in turn that gave them more choices of food in greater quantities, which gave them more free time to think and play instead of looking for food. Fire kept them warm and was the center of social gatherings and events where they exchanged information and strengthened their bonds. All this new information and stimulation of the brain didn't go unnoticed. In less than two thousand generations the brain doubled in size and its capacity to learn and understand. The human brain became about the same size that today's human brain is around four thousand years ago.

At that time, humans occupied all the continents on Earth, except for Antarctica, for many generations. They spoke many different languages and dialogues. Their appearance, size, brain capacity and behavior were noticeably diverse to mirror their diverse locations, environments, food supplies and gene pools.

All those early human groups had a leader. They couldn't survive without a leader. Without a leader the group would disintegrate and could become easy prey to predators and rival groups, and their genes might come to a dead end. If something happened to the leader, there were always other adults to take his place. The urge to lead or be led comes from the group of genes that pass from one generation to the next, and they are present in the DNA of all social animals and in animals that survive by grouping.

Many groups of humans lived in greatly distant locations from each other for many generations without exchanging communication and information with other groups. Most groups stayed within a fifty mile radius for many generations without meeting any other humans. It was dangerous to walk alone for fear of predators and the unknown. So, these groups didn't flourish mentally and physically as some of the larger groups or close-knit numbers of groups that could communicate and pass more information among themselves. The latter groups had more diversity and a better gene pool, and they stimulated their brains close to their full capacity.

They started to draw signs as a way to send messages or to keep information. (Nature repeats itself: the human discovered billions of years after the first gene that the shape of objects contains information, or a message, and can be preserved.) They domesticated the horse and used it to discover new territories and to relay messages to faraway groups. They already tried to understand the forces of nature and could think in abstract ways. They built statues and tall structures to send messages, to please their egos, to mark and preserve their identities and hopefully to please the forces behind nature as they understood them.

The human brain needs constant stimulation, especially in the early years of development when the memory cells are not developed to their full capacity and the brain connections aren't defined. The brain is more flexible and much easier to mold in early childhood and becomes stiffer as it gets old, just like the body.

The human brain is the most complex structure in nature. It has its own closed circuit of blood that supplies it with oxygen, carbohydrates and other necessities to keep it in a certain order and temperature that is the most beneficial to the chemical reactions that take place within it. When the body is resting, the human brain uses the most energy in the body, with the exception of the heart. The brain creates and disintegrates molecules in a split second with an electron's pulse. It has the ability to think like no other creature on Earth, and it is large because it holds more matter than any other brain. This matter, in a certain order that the genes built with the right connections and movements, gave rise to thinking.

We share over ninety-six percent of our genes with our closest relatives, the chimpanzees, and this less than four percent represents about twelve hundred genes. These twelve hundred genes make us look different and think differently from the chimps, and most of these different genes are involved with our brain size and function. The human brain has more memory cells than the brain of any other animal. The human can take the information that is stored in the brain and think in the abstract without receiving any new information, just by picking out different memories to make a vision or a thought. The advantage of a brain in the evolution of life is clear. The brain gathers information through the body's sensors and tries to respond to new situations by ordering the body what to do next. In other words, information is taken from the past, even a split second from the past, and the brain orders the body to respond in the future, even a split second in the future. So, the brain is able to see the future and respond to it by calculating and comparing information from the past that is stored in its memory cells to the information it just received through its sensors, and it accordingly orders the body to make its next move.

A bigger brain with more memory power and better reading and processing centers for information could see the future faster, with less new information coming in, or it could see much farther into the future, with the same information coming in, and by doing so could give the life form that possesses it a big advantage for survival. That's what the human brain has. The conscious part of the brain is the biggest and most powerful brain in the animal kingdom. It is what makes humans human. They can see much farther into the future. They can build tools and weapons as an extension of their bodies in much less time than it takes nature by way of genes, and by doing so, they can accelerate their own evolution.

There is nothing holy in the brain and its functions, and nothing in it is made from a different dimension. It is made from matter. It has four dimensions just like any other matter structure and it obeys the laws of physics of our universe. Genes, which are also matter and are nature-made, build the molecules in the brain in a certain order and pattern to accomplish results that last for a long time, or in other words, to survive the same pattern for a long time. Genes do that by trial and error with no touch of a divine force.

The human brain receives information from the body's sensors as electron pulses through the nervous system, or as a stream of molecules through the blood supply, or by both mechanisms, and creates new molecules through the chemical reactions of this new information inside the brain. The reflection, or reading, of these new molecules and the old molecules in the brain, and the reaction between them at any given moment, gives rise to a chain reaction, to thoughts and urges that order the body's movement and action by sending electrons through the nervous system or molecules through the blood stream (or both) to respond to the new information. It also carries information in its collective memory that passes from one generation to the next, informing us how to behave and act in a given situation without having any experiences. Everything is geared for survival, the preservation of identity and the passing on of genes.

The brain creates the outside universe inside of itself. It is a universe as big as understanding, but is still a whole universe. But, to survive the outside universe, the brain needs to be in touch with outside reality and to quickly respond to changing conditions. How fast and what kind of response this is depends on the brain's ability to process information by making new molecules, fusing and disintegrating them.

The creation and disintegration of molecules in the brain always involves the movement of electrons that the molecules share. So, electron movement is a must for the operation of the brain. With an electronic scanner we can see electrical activity in different parts of the brain that are responsible for different activities and information processing. If we do not see electric activity in the region of the brain where it is supposed to be, that region is dead or is not receiving any signals. If there is no electrical activity in the whole brain, the brain is dead and its universe has died with it. But the body attached to the dead brain can still survive with outside help.

So, the healthy brain works by producing molecules from the information it receives through the body's sensors with the help of the raw material it possesses and the energy provided by carbohydrates, which allows interaction of new molecules with the molecules that come from its memory cells (the short-term, long-term or collective memory cells) and creates a molecule that represents a new information unit by way of its molecular structure. The chemical reflection, or the reading, of this information unit by the brain, also by chemical reactions, represents the processes of thought, imagination and feeling.

Sometimes the message (or the reading) is not very clear because the new information molecule changes its structure repeatedly and the brain cannot read it clearly. This usually happens when the brain receives new information and creates new molecules out of it, and these new molecules interact with molecules that came from short-term, long-term or collective memory at the same time or within a close period of time. The brain then gets a mixed reading and mixed information from these new molecules. It tries to put order in them by breaking each molecule down into smaller ones, which it does through interaction with electrons and other molecules from the memory cells. It then acts once the message or the information is clear.

The brain gets a huge amount of information at any given moment and it does not have the capacity or the energy to transfer all this new information into molecules. Therefore, some information goes unnoticed and does not have any record in the brain. The brain makes molecules out of things of which it is aware, things that are needed for survival or things that stimulate its senses, like fear, joy or curiosity. These new information molecules are sent for storage to the short-term memory cells, and depending on their importance can last from a few seconds to a few days. If they are not used they disintegrate and the information they carry vanishes. (If you do not use it, you lose it. Sound familiar?). The molecule that carried that information disintegrates and its raw materials are reused in the never ending cycle of the creation and disintegration of molecules in the living brain.

In an emergency situation, during a life threatening or very important event, the brain shuts off all or most of the information coming to it and does not transfer all the incoming information into molecules. Instead, it concentrates only on the information that represents this important or emergency event to facilitate a quick response. New molecules that were created to represent the information of the last important/emergency event usually go to the long-term memory cells to provide knowledge of what to do in a similar situation, how to survive it, or how to avoid the same situation.

Repeated information also goes to long-term memory as you learn how to walk, swim, speak or write. Some of these molecules disintegrate after long time periods if they are not used, but most become immortal as long as these long-term memory cells are functioning and have good connection to the rest of the brain. Upon first-time events, when the brain makes a molecule out of new information it has just received, it usually stores that new molecule in its long-term memory cells.

The human brain develops in its mother's womb a few months after conception, together with the body's sensors and nerves, and is already in action at that point. Most of the information it receives is repeated information, like the mother's heart rhythm, her movement and voice, and the surrounding temperature, and some of it is new information, such as loud noises or sudden movements. But most of all, it remembers the shocking event of birth, when the newborn baby comes out of its mother's womb and sees the blinding light outside for the first time. The light is bright white.

All this repeated and new information goes to the baby's brain and forms the basis of its ever growing long-term memory cells. This information will materialize only if the human has a brain disease, is under a strong influence of drugs, or because of a near-death experience during which he can hear, feel and mostly visualize his life going backward all the way to his mother's womb, where his long-term memory started. He will see all the things that influenced his life: his parents, his family, his close friends, his childhood, and all the beautiful sceneries he had long forgotten, all the way down to the first light of coming out of his mother's womb, and then farther back to the weightlessness, warmth and security inside his mother's womb, and finally, to his mother's heart rhythm.

All life forms either suspend life for a short or long period of time by sleeping or hibernation. It is essential for their survival to wait for better conditions for their own livelihood, to conserve energy, or to rest and heal. The suspension of life by organisms such as bacteria or plants is triggered by not having the right environmental conditions, such as water, food, temperature or light. The organism suspends most of its activities and comes back to life when the conditions are right. A few plants and animals have internal chemical clocks to suspend life and play dead in order to outlast predators that prey on them. Some life forms go into hibernation that is triggered by external conditions such as the changing seasons, and also by an internal clock. They wake up when the conditions are right for them, or by the timing of their internal clocks. In hibernation, the body suspends some of its activity and slows the rest.

Humans and some animals go into a sleep cycle that is triggered by an internal clock regardless of outside conditions, even though the internal clock cycle is triggered by the cycles of outside conditions (day and night) in the first place. Sleep helps animals conserve energy by resting and healing their bodies, but is mainly necessary to rest their brains. The brain is normally very active. It makes molecules out of the information it receives from its sensors and from the body, and moves them around to different parts of the brain. It creates and disintegrates a huge amount of molecules and has the most electrical activity in the body.

The brain can be damaged if it gets too much information over a long period of time. That is the reason why the brain will not work all of its different departments close to one hundred percent of their capacity at the same time. This would cause it to overheat, short-circuit and have irreparable damage. It is like working all the muscles in your body to one hundred percent of their capacity. This cannot be done for more than a few seconds. So the brain has to have a break on a daily basis in the form of sleep, when it suspends most of its activity and heals itself. Sleep helps the body and the brain to correct and optimize their chemical imbalances. Lack of sleep will cause the body's immune system to be weakened. It will create nervousness and short temper, impair judgment, and cause overeating (the conscious mind gives easily to subconscious pressure).

Sleep is very important to a brain, and especially for the large and complex human brain. If the brain is deprived of sleep for more than a few days, it will be damaged beyond repair. Total sleep deprivation for more than a few days will also cause the brain to hallucinate, to have daydreams, to disconnect from reality, and will finally lead to death.

The young brain of a baby or a small child needs more sleep because their brains are working very hard due to all the new stimulation and current molding process. During sleep the brain suspends most of its activity and is not making molecules out of the information it receives from its body and its sensors, especially the eyes. The eyes are shut because the information that comes through the eyes takes the most matter-energy to convert to information molecules. Therefore, the brain doesn't have any recollection of what's going on around it while it sleeps and will likewise have no memory of what happened when it gets up from total sleep. However, the brain does take some information from its sensors and body, such as a loud noise or a touch, since it still oversees the survival of the body for which it is responsible.

While asleep, the volume of information coming into the brain depends on the sleeping stage and varies from brain to brain. When the brain is asleep its activity slows down. It tries to heal itself by repairing connections and releasing pressure points, which are information molecules that have less or more electrons than they are supposed to have. These molecules are unstable and the brain tries to move them around to place them in the right place and order. These pressure points are molecules that contain information and could come from the vast pool of information that short-term, long-term and collective memory cells possess, or from new molecules that are wandering around because they were left over from before the brain went into sleep.

The reading of these information molecules while asleep constitutes a dream. The brain brings the information that these molecules hold back to life. Most of the time the reading of these molecules that come seemingly at random from all the memory storage does not make sense. It is weird and out of order. The illusion of time is messed up because the timeline of events is disorganized, and it isn't clear which events came before or after others. This happens especially when there is a strong influence or pressure from collective memory, which holds the information of the human brain throughout the evolution of many generations of many different species up to this point in time.

There are lots of dreams involving pressure points that mainly come from short-term and long-term memory cells, with a small influence of collective memory cells. They are triggered by new molecules of new information that the brain just received from its body sensors while it was asleep, such as external temperature, noise or light. These dreams will make more sense, are easier to understand, and will involve recent events. So the brain dreams all the time even when it is awake, but most of the dreams are forgotten since they are not stored in memory cells and especially not in long-term memory cells.

When an individual has a problem or tries to remember something, sometimes it creates a pressure point in his/her brain, and when the brain goes to sleep it tries to solve this problem to the best of its ability by moving molecules around. When the human awakens, the problem is solved. The brain picks up the right molecules in the right order to the best of its ability, and the reading of these molecules gives the answer (which does not have to be the absolute right answer).

For example, if you put pressure on your brain, let's say to find out the right combination of numbers for the next day's lottery, it will come up with a number to the best of its ability. But since it does not have all the information about the huge number of interactions that led to a lottery number, it will come up with a random number which has a chance to win that is dependent on the odds of winning the lottery combination. Sometimes it will come up with the right combination, but most of the time it won't, because it is only a random number. The brain cannot see the right future without having all the correct information and the interactions that can lead to an event or the positioning of matter in the future. The brain can see the future only through calculations of the right information and interactions that lead it to predict future events, and the brain can do that even in its sleep.

The brain is completely sealed from the outside and its connection to the body is only through tissue that attaches it to the body and the blood vessels that supply it with nutrients, energy and some information molecules. The blood vessels also remove waste materials and take information molecules back to the body. The brain is also connected to the body's nervous system, which transmits information back and forth by way of electrons.

The brain tries to recreate the outside universe inside of itself by making molecules. These molecules carry a huge amount of information because they are made from many atoms and contain information in three dimensions. In other words, the position of the atoms in the molecule carries information. The brain holds a huge amount of information molecules, plus a huge amount of connections, storage and processing centers, and that's why it needs to be large. And the human brain is the largest in the animal kingdom. Its capacity is the biggest and it can see much farther into the future when it involves the information stored in the short-term and long-term memory cells and the ever present influence of collective memory. The processing of this information in addition to the new information coming in, gives rise to awareness and consciousness of what is going on outside of the brain.

In reality, the brain is disconnected from the outside. It only reads the molecules that represent information coming from the outside or the molecules that come from its memories. So, the brain is really at the mercy of those molecules. These molecules represent information by the way they are structured, and if their structure is influenced by other things it changes. Therefore, the information it carries is different from the reality outside of the brain, but the brain perceives this as reality. The brain cannot see or hear what's going on outside of itself. It can only read information through the different structures of the molecules that represent information, or more correctly, through the chain reaction that the position of the molecules creates. If the structure of a new molecule that was just created to represent reality changes because it is influenced by other molecules coming from memory cells, or by drug molecules, or by a pressure point floating around, it will change the brain's perception of reality.

There are many chemical reactions in the brain, and with the right connections those chemical reactions make the brain function the way it does.

There are some drugs (or foods) that can accelerate or slow down the brain's action and therefore its response time. There are some drugs (such as LSD) that react with the new molecules of information coming in from the sensors and can twist them to look different from the reality outside, or can even disconnect all information coming in and make their own molecules. And that's what the brain sees, hears and feels. It will have its own universe within, disconnected from the outside universe.

There are things other than drugs that can create the same effect inside the brain. A strong magnetic or electric field will bombard the brain with electrons, and those electrons will interfere in the process of building new molecules since the electrons are really what hold the atoms together to form a molecule. If the molecule has too many or too few electrons it becomes unstable and easily reacts with other atoms and molecules to create new molecules with different structures that represent different information. The same thing can happen even without the bombardment of electrons from the outside.

In very few people the new information coming in can sometimes be twisted by their own pressure points. These are molecules the brain makes when it is anticipating something or waiting for something to appear for a long time. These pressure points can also be created by collective memory, which has a very strong influence on the brain and can create a new chain of molecules that represents a wish or an urge that comes from collective memory, which creates a reality that is disconnected from the outside reality, but is nevertheless a reality (like you holding this book).

Another phenomenon of creating and disintegrating molecules is that they produce electron waves, and the brain produces and disintegrates a huge amount of molecules in a split second. These waves of electrons, with their very low voltage, can be picked up by the electronic sensors that some animals possess and also sometimes by the human brain as a telepathic sense. You could feel that there is a brain nearby without seeing it and very rarely the brain could pick up some of the thoughts of another brain nearby, if they were able to synchronize their waves.

Today the human brain has reached its full capacity and people use their brains more than ever before. If we continue to use our brains to their full capacity they will continue to grow in coming generations. It is the way of our body parts. The more we use them, the bigger they become in subsequent generations, whereas if we do not use them, we lose them in the following generations.

23. The Subconscious (The Collective Memory)

The subconscious, which also contains collective memory as an integrated part of the subconscious, is the most influential part of the brain. It is at the center of the brain and the rest of the brain developed from it. It contains genes that are passed on from previous generations, which hold information on how to develop the brain and how it will react. The subconscious can create information molecules of past experiences from many generations ago, through the different branches of a life form all the way back to the original brain that evolved about one billion years ago.

Once the simple brain developed, it gradually became bigger and held more information and experiences of the collective memory of many generations. It became more complex with forward movement of time and with changing conditions. It developed the long-term and short-term memory cells to hold information temporarily in order to save energy and space. These long-term and short-term memory cells can hold information molecules and disintegrate them if they are not needed or used to make room for new information molecules to be stored. The disintegrated information molecules are recycled to make new information molecules. These new long-term and short-term memory cells gave rise to consciousness, the awareness of what's going on around us, and later on gave rise to imagination.

With the passing of time the subconscious got bigger by adding more genes. Some of the long-term memory cells, which are really receptors, became genes together with the information molecules they held to become immortalized in DNA, and passed on from one generation to the next as part of the subconscious mind. That is the reason that different species behave in their own particular ways, and why it is possible to experience or feel events that your mother or father experienced, or that people only a few generations ago experienced. It will come to you as deep knowledge or familiarity with things that you are experiencing. It is as though it happened to you before even though it never did, like when you go somewhere and the place looks familiar, or you get certain feelings about the place and you never have been there before.

But this is only a small aspect of collective memory and of the subconscious in general. We have to remember that the primitive brain developed to give a survival advantage to the species that possessed it, to control behavior and movement of the body in different situations and to pass that knowledge from one generation to the next through the genes in DNA.

The subconscious controls the conscious mind through feelings of deep knowledge or unexplained urges, through pain, fear and joy, the feelings of nervousness, anger or madness, the feelings of happiness or satisfaction, love and affection, and all the emotions involved.

Unlike the conscious mind, the subconscious mind does not predict the future or the long-term effects of its behavior. It does everything right away without thinking, as an instinctive reaction to a given situation without a time lag. In contrast, the conscious mind has to control subconscious urges by taking the time to think. The conscious mind thinks by comparing molecules from its memory cells and picking the right one for the new situation given, and all that action takes time. It compares memories in order to predict what will happen in the future if the urge that comes from the subconscious is fulfilled.

But since the conscious mind is an extension of the subconscious (the primitive brain), it has the same goals of giving the body that possesses it a better chance to survive by seeing the future, by saving energy, or by gaining the most energy for the body by using the least energy. For example, an experienced predator will not chase a healthy animal from a long distance away even though its subconscious tells it to eat by producing hunger pains. The conscious mind will calculate the distance and the ability of its prey to resist getting caught or to fight back, and therefore will chase only a nearby animal that will give the least resistance. The conscious mind's reward will be the feeling of joy, happiness and satisfaction when this predator has achieved the goal. If the predator's conscious mind is too weak or inexperienced, it will chase far away animals without success, resulting in the waste of precious energy or in a fatal fight with a stronger animal.

So, the subconscious controls the conscious mind by making feelings of joy and happiness, fear and pain, and the urge to pass on genes. You cannot escape these feelings unless you satisfy them in various ways. Some drugs can mimic the same molecule that the subconscious mind produces and can give you the same feeling of joy and happiness. Some drugs can block the effects of fear and pain coming from the subconscious, which disconnects the conscious mind from its subconscious and from the body to give you freedom in thinking and imagination. Many people cannot resist the temptation of using drugs to overcome these obstacles of their bodies. But nature intended to connect body and brain to face reality, so using drugs will cost a price. Because the molecular structure of drugs is not exactly like that which the brain provides, and neither are the quantity and the timing, using them may result in some bad side-effects.

The subconscious brain is created by many genes, and some of these genes store information of past experiences and memories. This is where the name "collective memory" came from. These genes produce molecules under certain conditions and the way these molecules interact in the brain makes each species behave the way it does.

The main goal of all species is to pass on their genes to the next generation and to preserve their structure, their identity as a species, which is the meaning of life. But to do that, each individual from each species has to survive, and that's where the power of the subconscious or of collective memory comes in. It gives the individual the knowledge of its own species' past experiences up to the present time, giving it the behavior to survive as an individual and as part of a group. Sometimes this can lead to an individual being sacrificed for the benefit of the group.

For example, let's start with the smallest group, consisting of a mother and her offspring. Through the impulse of love and devotion, the mother's subconscious tells her to take care of her offspring. She will spend most of her time taking care of her offspring, just like a slave, providing food and shelter. And she will give her own life to protect them as an apparent act of madness when the subconscious takes total control. There is no fear impulse; she will fight a much stronger enemy against the odds if she feels they might harm her offspring, and even if she gets hurt or killed the message is clear not to mess with the offspring when the mother is around. She will fight and can inflict injuries, but it goes beyond that: it gives that species as a whole a better chance for survival by having more offspring that reach adulthood.

So all these feelings for her offspring and the knowledge to take care of them come from her subconscious brain without thinking. Her reward will be the impulse of happiness, joy and satisfaction that also comes from the subconscious mind when everything goes well. She satisfies the urges of her genes and the genes have another chance to copy themselves more or less the same way, to beat Father Time and to preserve their identity, their unique structure.

On the other hand, the young offspring is controlled by his subconscious mind only, since his conscious mind is only in the early stages of development, gathering information. His subconscious mind will instruct him how to look for his mother's milk and cry when he needs something, and later on in his life he will follow and hang on to his mother for safety and comfort. The forces that drive him will also come from his subconscious as an impulse of fear and anxiety whenever he does not see his mother. His mother will act as his first leader; he will follow her and trust her to take care of him.

When he grows up he will follow the leader of his group, or he may become the leader himself. This impulse to follow the leader comes from the subconscious. The subconscious has a group of genes that are responsible for the pulse or urge to follow the leader, or anything that resembles a leader, blindly and without thinking. The urge to lead also comes from the subconscious and sometimes it can be very strong depending on the conditions.

Humans have always lived in groups, but the genes that are responsible for this urge to lead and to be led as a way to bond a number of people together into a bigger and stronger group were created many millions of years ago even before the mammal family was created. These genes gave a better chance of survival to any small and weak individual by forming a group of individuals that could act as one larger and stronger force. This larger group could deal with predators in various ways, find food, communicate, and improve the gene pool in a way that is only possible for a large number of individuals in a small place. So the genes that hold information and provide this knowledge had a long history and time to become more complex and to become embedded in many different species. The influences of these genes in some species of the mammal family are very strong and will create the urge to follow the leader anywhere, even to death. Some animals will follow anything that resembles a leader, even if he is from a different species, because their subconscious gives them a strong urge to follow and obey.

Let's take dogs as an example. Wild dogs always live in close-knit groups and they are very successful when they form a group, whereas a single dog in the wild has a very small chance to survive alone for a long time. A group of wild dogs will have many dogs and one leader. They closely watch the leader and try to understand his orders because their subconscious gives them a strong urge to follow and satisfy him. But since some dogs have been domesticated a long time ago by humans, these domesticated dogs see the man as their leader and will follow him and obey his orders. In some dogs the urge to follow the leader is so strong that they will stop eating if they are ordered to do so, or cannot function if the leader is not around. They will feel sadness, hopelessness and like part of them is missing.

Another benefit of the genes that give you the urge to follow a leader is that they can save you energy by allowing you not to think. You just follow and obey the leader and he does the thinking for you. The leader is looking out for your own good and you trust him. The success of these genes results in close-knit and successful groups.

To varying degrees of intensity, humans as social animals are also influenced by these follow-the-leader genes in their subconscious. This urge to follow the leader can manifest itself in many different ways. For example, when you drive and do not know where you are going on the road you will follow the car in front of you, or you will take the lane with the most cars in it. Or when you are on the road driving without thinking, you will follow the car in front of you and drive like the autopilot is engaged. The autopilot is the subconscious, with the influence and the urge to follow the leader, and the leader in this case is the car in front of you. Without these follow-the-leader genes, humans would not be able to create state government structures or the regimented structure of the army. The structures of all these institutes are created by following the leader and the hierarchies that result.

Without these follow-the-leader genes a herd of goats wouldn't stick together and wouldn't follow their Shepherd.

There are many "dark" sides of the subconscious, and also "good" sides. Everything is geared for survival and the preservation of identity for the individual, the group, and the species as a whole. I will mention some of the "bad" influences and urges that come from the subconscious, and then I will explain them, and afterward, I will mention the "good" influences and explain them. First, I want to say that there is no absolute "good" or "bad" in nature. Everything is only relative. Everything has a reason, and what is bad in one place could be good in another. Some of the "bad" influences are fear, pain, deception, mistrust, overeating, the urge to have sex, racism, madness and killing.

Fear is the influence of the subconscious on the conscious mind to avoid places or situations that can hurt or kill you, such as high places where you can fall and hurt yourself or dark places where you cannot see whether you are alone and from where the danger will come if it does. Fear will prevent you from engaging a strong enemy. The fear could be so strong it could paralyze you, and in a way it is another tool for survival. You could freeze and be undetectable to your enemy, or play dead and not give a reason to be attacked.

Fear can be overcome by drugs that block the fear molecules that the subconscious produces, or in emergency situations when there are different aspects of the subconscious taking control, or also by the conscious mind after it accumulates knowledge through practice to learn how to deal with different situations. Fear is an important tool for survival. Fear prevents people from engaging in dangerous situations, as the subconscious sees or is programmed to recognize them. A fearless person will take too many chances and engage in dangerous situations, so that his chances of survival are very slim and he might not pass his genes to the next generation. On the other hand, a fearful person won't take any chances and will be afraid of his own shadow and might not pass his genes to the next generation either. The intensity of fear is different from person to person and from time to time.

Pain is the message from the subconscious to the conscious mind that there is a problem with some of the body parts where the pain originates. It is a way to connect the body to the mind and force the mind to take care of the problem. If the problem is severe, the pain will be more intense in order to incite immediate attention so the problem can be taken care of by not using that body part so that it can heal itself. Pain can be overcome by drugs that can block the pain message or by the subconscious in extreme situations when the subconscious needs to save the life of the individual or the group to which the individual belongs. For example, even if you lose a leg or a hand while fighting in a war, you won't feel any pain. On the other hand, it's possible to sometimes feel pain from limbs that you already lost, which means that pain comes from the subconscious. Pain is important for survival, because without pain you would be disconnected from the body and could get hurt easily. The feeling of pain is different from human to human and from time to time, and it is influenced by different conditions.

Deception is when the subconscious tells the conscious mind to deceive, to change its behavior in order to gain something with less energy than it would otherwise spend. The genes that are responsible for deception are billions of years old, and in fact, these genes deceived time from the moment they were created, and by making copies of themselves deceived nature by changing their structure through evolution to adapt to different conditions.

Even a virus, which does not have a brain, changes its shell so that it won't be recognized by the body it invades. It is the nature of genes to change, to adapt and to deceive, and the subconscious reflects this ability. Deception has many different forms in different species. For example, when a non-poisonous snake has the same color pattern and look as a poisonous snake, he is deceiving his predators. When animals blend the color of their skin or fur to their surroundings, they are trying to deceive their prey or their predators.

The genes that are responsible for the urge in the subconscious to deceive became very complex with the passing of time. They came to control the behavior of the species with the knowledge and experience of many previous generations in order to improve the chance of survival. For example, a tiger lays low and hides behind tall grass to ambush its prey, or walks low and very quiet against the wind to catch its prey by deceiving the prey's vision, smell and hearing sensors. In humans, the urge to deceive is ever present in order to gain something or to be something that you are not. Actors make a living out of this talent of being somebody else because they also have a strong conscious ability to control this urge. People without strong control of the conscious mind lie for no apparent reason whenever a strong urge to deceive comes from their subconscious.

Mistrust also comes from the subconscious, from the knowledge that there is a lot of deception around you and that things are not what they appear to be. So if you want to survive, you have to check everything with care, especially things that look too good to be true. The subconscious knows that and gives the urge for caution and mistrust. A deer will sniff the air around him all the time and use his ears and eyes for the first sign of trouble, because he was born with the knowledge to be cautious.

In humans this urge can vary. Too little will cause him to fall victim while too much will make him nervous and frustrated. This urge is more likely to kick in when you face a strong force that can hurt you, than when you face a weak force or person. This urge to mistrust is the main driving force for conspiracy theories among humans. This urge can be controlled by drugs and by a strong conscious mind.

Overeating is a basic urge because there is no life without consuming food. The strong urge to eat comes from the subconscious: from the collective memories of many past generations when food wasn't available all the time. In earlier times, when food was available you gorged on it until your stomach was full, or food was running out and you had to eat quickly before somebody else got it. If you had a choice you would go for rich foods with a high caloric and protein content. The intensity of this urge is different from human to human and depends on other conditions such as fear or anxiety, when the conscious mind can no longer control and discipline the subconscious urge to eat like a pig. This urge can be controlled by drugs or by a strong conscious mind.

The urge to pass on your genes is basic and very strong because there is no life if you cannot continue the cycle of reproduction. In humans, reproduction is achieved through sex between a man and a woman. Collective memory is different in a man and a woman because of differences in physiology and therefore in their roles in the reproduction of a new generation. Also, because males and females had different experiences throughout evolution, their collective memory is different.

Let's start with the male of millions of years ago. The adult male had a strong urge to pass on his genes, which means he had a strong urge to have sex regardless of whether this would result in offspring, since the subconscious cannot see the future like the conscious mind does. Not every attempt to pass on his genes resulted in offspring. So the male would attempt to have sex many times with as many females as possible to release the pressure of this urge. Nature works by way of huge numbers in order to have success that will result in new life (sperm count). Therefore, the male would have sex any time he had the chance to. The urge was triggered every time he saw a female that behaved or smelled like she was in heat or ready for sex. But he wasn't alone; there were many other males around him that tried to do the same thing, and he had to fight his way to the female. So, the stronger male had a better chance to pass on his genes. But to make sure he was the father he had to watch the female to make sure she did not have sex with other males. The weaker males had to deceive the stronger males or abduct the females by force or by offering them food (or promises of marriage) in order to have sex and pass on their own genes to preserve their identities.

Every time a group of humans invaded another group's territory they killed everybody except for the young females, who they raped in order to pass on their genes (their identities) through them. So, the passing of genes throughout the history of mankind involved a lot of violence, and these memories of violent sex are embedded in the collective memory of every human being. This resides in the dark side of a man's or woman's subconscious, and if it is not controlled by drugs or by a strong conscious mind it can come out to haunt us (as rape, torture and murder), which it does especially in times of war or chaos when the conscious mind knows there is a very small chance it will have to pay for its crime. The conscious mind is, after all, a tool for the subconscious to see the future.

As with any other species, humans share the same genes as the opposite sex, but not all their genes are activated, or activated to their full capacity, and that makes them look different and function differently from the opposite sex. But regardless of physical appearance, the feeling and the behavior of males and females comes from a group of genes in their subconscious mind. In other words, it is possible to have a body like a female and still behave, think and feel like a male, and vice versa. (So, in a twisted way of nature, we are all homosexual. We are all the same, male and female.)

A female's subconscious will give her an urge to highlight her sexuality around the time she turns twelve years old. She will alter her behavior, makeup and clothing habits, and seek a strong and violent male who can protect her from other males as well as protect and provide for her offspring. We can sometimes see this unexplained urge in a woman who tries to get together with a big strong man who looks like a caveman, or with a convicted murderer or rapist in jail. Basically, the subconscious mind of most females will influence them to seek older, powerful males, in contrast to most males whose subconscious will influence them to seek weak, young females.

Some people can only get aroused sexually through pain and violent activities because their subconscious (collective memory) connects sex with pain and violence.

The urge to pass on genes is basic and very strong. Humans do it through sex, and the deprivation of sex will result in the rise of other urges such as killing, violence, short temper and rage. The urge for sex can be controlled by a strong conscious mind and also by drugs.

Racism comes from the urge of every life form to preserve its genetic identity and therefore resists diversity. After all, the gene's structure is an information unit, and if the structure changes, so does the information. This is the reason why different species do not mix genetically. This urge is basic in humans also. Every human being wants to make sure that his or her genes are duplicated and that their offspring look and behave like them. This is true for individuals and also for a group within the same species. The group likes to see that all of its individuals look the same and behave the same, and this gave rise to racism in humans.

Madness is not an urge that comes from the subconscious, but it occurs when the subconscious takes total control by overriding the conscious mind. It happens when the conscious mind is damaged, underdeveloped or when there is an emergency situation and the subconscious takes total control. The subconscious mind is composed of all the basic animal instincts and it behaves with no regard to the future or to anybody else. A baby will fulfill his needs anywhere and at any time because his conscious mind has not yet developed to take control of a situation, but if an adult human being fulfills his needs anywhere at any time it means that his conscious mind is not in control and that it gave up control to his subconscious mind. In emergency situations, when the human life is on the line, the subconscious mind can take total control, and the human will fight using all his body parts and anything else he can put his hands on. He would look like the animal that he is in the subconscious part of his being. Under these circumstances the conscious mind will not function and the human will lose his ability to speak clearly and to see the future consequences of his actions. When any animal or human gets cornered by a stronger force, his subconscious mind likewise takes total control and he will fight or attack even though the odds are against him.

Killing is also an urge that comes from the subconscious mind. It is embedded over many generations in the collective memory of many different species. Killing is a basic urge. It is a necessity for survival: to kill for food or to get rid of the competition, to gain territory or just to preserve identity. The urge to kill is embedded in the subconscious mind of every human being and its intensity is different from person to person and from time to time. It can be triggered for many different reasons, and like many other urges, it can be controlled by drugs or a strong conscious mind.

The good sides of the subconscious mind are the urges to help and the feelings of love, happiness and satisfaction. The urge to help is very strong in humans as a species and is embedded in the collective memory of many species as a way to help their offspring to be independent. This way, the offspring can continue to pass on their genetic identity to the next generation. As social animals, humans live in close-knit groups, and by taking care of everybody in their group they can better their chances for survival. Sometimes the individual will put his life on the line to save and help others in his group or anyone with whom he identifies. This urge is in every human and its intensity is different from person to person and from time to time.

As I mentioned before, there is no real differentiation in nature between "good" and "bad". So the urge to help could even come when you help to kill others, if by killing you help to save your own group or anything with which you identify, just like soldiers in a war.

The feeling of love is a strong urge also: the urge to be close to someone, to help them, to provide for them, to tolerate their negative effects, or to raise their helpless offspring. This love urge is in many species. It helps their bonding as a group, and it is well developed in humans. For this reason, the human race flourishes. The intensity of love is different from person to person and from time to time.

The feelings of happiness, joy, pleasure and satisfaction come also from the subconscious mind when everything is going well and all your subconscious urges are satisfied. These feelings occur when you satisfy the purpose of the design of your body made by your genes, of which the subconscious and collective memory are made, like the good feeling you get from watching lakes and waterfalls surrounded by trees and luscious greens. This good feeling arises because you are looking at scenery to which we have been accustomed for millions of years, which is the scenery (environment) that gave us the best chances for survival. So, the good feeling we get from watching that scenery comes from deep within collective memory. It also happens when we see the Sun, or any other light source, because we have been day creatures for millions of years and it is good to see the light after the darkness and the danger of the night. Whereas dry desert scenery might look beautiful to us, but we do not get good vibes from it because this environment is very challenging for survival. When you satisfy the strong urges of survival or the urge to pass on your genes and see the offspring doing well, the intensity of happiness and satisfaction is the greatest.

You can also get satisfaction through stimulation of your conscious brain by playing games, reading, fulfilling curiosity, watching a show, listening to music or telling jokes, all of which stimulate the brain because they show new angles to look at different things. The genes design the conscious brain, especially the big human brain, and allow it to be stimulated and to expand. The intensity of happiness is different from person to person and from time to time and depends on the genetic makeup of each person, and all those feelings can be controlled by drugs.

If you pay attention, you will notice that all feelings can be created, enhanced or diminished by drugs, unless there is physical damage to the brain due to accident, disease, genetic disorder or old age. This can only be corrected by genetic therapy or surgery to connect or replace damaged tissues. All this means is that a feeling is a reflection of certain orders of matter in the right place at the right time and that without this matter in a certain order and with the right physical connection there would no feelings, urges, thinking or imagination.

The urges that come from the subconscious are very strong. Every person is controlled by them and is capable of fulfilling them for better or worse ("good" or "bad"), regardless of his or her appearance or background.

Our subconscious mind gives us the deepest knowledge, the deepest feelings, and the morals to know what is right or wrong for us as individuals and as part of a group. For example, our subconscious will tell us through deep feelings that it is morally wrong to invade the territories of other individuals or groups and/or to use their belongings, or to pass on our genes through other people that do not belong to us.

The subconscious will also tell you through deep feelings that it is morally right to help the needy, to take care of the young ones and to sacrifice the few for the many in the group and the few or the many for the leader of the group. The leader could be flesh and blood or an imaginary leader: the feelings that come from the subconscious are the same.

24. The Conscious Mind

Only the conscious part of the brain contains the power of thought and imagination. The conscious part of the brain is an extension of the primitive brain, or the subconscious brain, which was first developed about one billion years ago.

The subconscious brain started as a simple brain, as interconnected webs of nerves without power of thought or any imagination. These webs of nerves received information in the form of electron pulses coming from the body's sensors and going to the primitive brain. The brain reacted to this information immediately through a chemical chain reaction of these new electron pulses coming in and interacting with the genes in the brain. So, a new information molecule was created to provide knowledge of how to respond to a new situation (by way of this new molecular structure of information and the way it disintegrates). Therefore, the primitive brain didn't preserve any information or memory molecules, but was able to react appropriately to new situations.

With the passing of time and with changing conditions, this simple brain grew in size and complexity as a result of mutations and because it was used to its full capacity or beyond its capacity some of the time. Full capacity means one hundred percent of the volume, not one hundred percent of the time; it could be less than one percent of the time. For example, the more people use a dirt walkway in a field, the wider and more defined it will become. If the walkway is used to its full or over its capacity, some people will walk in different ways, not to take a shortcut necessarily, but to get away from the physical pressure of the crowd. They will make it wider and make new smaller parallel walkways nearby. On the other hand, if people hardly use the walkway, it will shrink. If they stop using it completely the walkway will disappear and blend with its surroundings.

On the same principle, body parts such as the muscles, blood vessels and nerves can become bigger and more complex or can become smaller, weaker and could disappear altogether out of the necessity to adapt to changing conditions and the pressure of everyday life. So, because the early brain was being used at or over its full capacity for some of the time, some of the new nerves that were created within it held a new information molecule at their dead-ends. This information molecule didn't disintegrate since it was protected by the nerve ending by mistake, and eventually these nerve endings, together with the information molecules, became memory cells, which held the new information molecules. The information molecule, just like the gene molecule, holds its unique information, or knowledge, with its unique structure. But the real beauty of these new memory cells is that they can copy the information molecule they hold in a split second if necessary and send that copy as electron pulses to the processing center, or they can disintegrate the information molecule and recycle its raw material to save space and energy.

The primitive brain also developed extra processing centers that are part of the conscious brain where the copies of information molecules meet and are involved in chemical reactions. The reading of these chemical reactions is what gives us the power of thought and imagination. Therefore, thought and imagination always involve information molecules from different memory cells and can be triggered by new information in the form of electrons or molecules that the conscious brain just received from its body sensors or from its subconscious.

So, imagination or thought always involves some information molecules coming from the memory cells. If the conscious brain picks up the right information molecule directly from its memory cells, the reaction is short and there is hardly any thinking or imagination involved, just like in the case of regular talking or walking for normal adult human beings (normal in nature means the majority). Walking is easy and involves hardly any thinking. The right information is always picked from the long-term memory cells since walking is a repeated behavior, and the reaction involves very few molecules. When a person is walking the brain does not think, "I will pick up the right leg first, make a step and then I will pick up my left leg and repeat that motion". It just happens automatically, because it is repeated behavior and the information molecules are available right away. Now, the story is different when there are obstacles in the way while walking. The more obstacles there are and the more complex they are, the more power of thinking and imagination is needed. The human will stop going forward and will think.

Thinking and imagination take time because they involve many information molecules from multiple memory cells, and it takes time and energy to get the result of what decision to execute next. The conscious brain has to pick up a few correct information molecules from different memory cells and compare them to the information molecules it receives from its sensors in real-time. This comparison gives the conscious brain some choices of how to react to a given situation, and these choices are imaginary. These imaginary choices are new information molecules that were created from the interaction of different memory molecules with or without new information molecules coming from the sensors or from the subconscious mind. Then the brain picks up one of these imaginary choices (the molecule that has the least electric activity, plus or minus electrons) and orders the body to move or behave accordingly. This is exactly the purpose of the conscious brain: to see the future, to imagine different case scenarios and to pick one that will give the answer for the best chance of survival or will take the least energy to implement.

Energy is very important for survival. If you waste more energy than you consume, you will not survive for long, and on the other hand, if you consume more energy than you waste, you can grow bigger, become more complex, live much longer and have a better chance to pass on your genes successfully. Through its power of imagination the conscious brain can give you the benefit of wasting less energy. It can imagine different scenarios in response to a given situation and implement one of them to the best of its ability, and by doing so can save you precious energy.

The conscious mind with its ability for reasoning and imagination evolved as an extension of the primitive brain about a hundred million years ago. Since then, the conscious part of the brain got bigger and more complex as it added memory cells and processing centers, and it is the most developed in humans.

But the conscious mind with its power of imagination is a prisoner. It depends on the body for energy and physical protection and it depends on the primitive brain for its survival instincts, which the subconscious imposes on the conscious by providing feelings and urges.

So, the conscious mind is not free. It is a tool for the primitive brain to see the future by way of imagination, and the primitive brain is a tool for the body to improve its chances for survival and to preserve its identity. The conscious brain does not have feelings. It gets them from the subconscious mind, which controls the conscious mind by producing a feeling of pain or pleasure, or anything in between those two feelings. (The line between pain and pleasure is very thin; it fluctuates and is different between different people at different times). And the conscious mind can enhance and glorify those feelings, and/or put a reason or a sense behind them.

The conscious mind is cold, without feeling, and resembles a computer in the way it operates. The computer, just like the conscious brain, has long-term and short-term memory cells and has a processing center to pick the right answer for each question. So the answer depends on its memory and the ability of the processing center to pick and connect the right information from its memory and then put the answer together. The electronic computer has come a long way in the last fifty years and can do amazing calculations much better than the human brain on a specific subject, but overall, the human brain does more different calculations and functions than any single computer can. However, this won't be the case for long, because the evolution of the computer is much faster than the human brain. So, in less than a hundred years the computer will surpass the power of the human brain and its ability to do different things at the same time.

Just like the conscious brain, the main purpose of the computer is to gather information and to see the future with its power of imagination. For example, you can enter information into a computer about the Moon's movement in the past and it will tell you the Moon's movement in the future, including solar and lunar eclipses many years into the future.

The conscious mind, just like the computer, is also a tool to save energy with its power to imagine. For example, you can enter information about a car into a computer and the computer can imagine (virtual reality) this car in different colors, sizes and shapes, and by doing this can tell how the car will perform and look, which saves a lot of energy that could be wasted on physically building different cars of different sizes, shapes and colors to find out how they perform and look.

And just like the conscious mind, the computer has no feeling or urges. If you tell the computer, "I will set you on fire," the computer will not have any fear and will not try to avoid being set on fire. Once it is set on fire, it will not try to avoid it, because it will not have any pain, any feelings, or urges to stay alive, even though it has the ability to see the future.

Luckily, our conscious brain is connected to our subconscious brain, which is connected to a body that gives it mobility and sensors. The sensors provide outside information to the conscious mind, and the subconscious mind provides the conscious mind with basic survival skills and the desire to live by way of feelings and urges.

25. The Connection between the Subconscious and Its Conscious

In this chapter we will discuss the connection of the subconscious to the conscious mind that contains the power of reasoning and imagination in humans. The human subconscious is unique and different from other species because it contains the collective memory of humans for many generations. Therefore the relationship between the conscious mind and its subconscious will be different between species and even between different groups and individuals within the same species, because they do not have the same exact memories in their conscious minds. This is due to the fact that nobody occupies the same space at the same time; they all have different experiences and therefore memories.

The same is true about the subconscious, which contains the collective memory of the species. As the species expands in numbers and occupies different locations, it breaks down into different groups of the same species, and as they face different conditions they have different experiences and hence memories, which are embedded in their subconscious as genes. These genes act as a memory cells in their collective memory, and pass from generation to generation. These genes contain information of past experiences and how to deal with them or behave in similar situations in the present time, by giving instructions to the conscious mind by way of urges and feelings on how to take care of a given situation. The subconscious mind is fixed. In other words, it is made out of genes that copy themselves in almost the same way from one generation to the next and cannot be changed, or change very little in the lifetime of the individual.

All humans have a subconscious, but they do not have the same gene structure in their DNA or the same number of genes that contain the same information in their subconscious. And therefore, all humans do not have the same urges and feelings, or with the same intensity, that the subconscious imposes on the conscious mind.

The human body is created by about thirty thousand genes, but the layout of these genes in DNA is different from person to person. Members of the human race share 99.9 percent of their genetic codes with other humans, which means we are all very similar. We are different by one-tenth of one percent, which represents about thirty genes. These thirty genes, plus the fact that the layout combination of the genes in DNA is different from person to person, make us unique and different from each other. The different combinations of the genes in our DNA make us look different from other humans, giving us different metabolic rates to digest our food, varying levels of ability to ward off diseases and different rates of aging. Some of these different genes will be responsible in building the brain and some will act as collective memory.

The subconscious is the primitive brain that evolved for almost a billion years to become bigger and more complex in its operations as a necessity to adapt to different conditions and give a better chance of survival by holding memories in the form of genes that stood the test of time and are still with us. These genes influence the conscious mind through urges, feelings and unexplained knowledge that all come from the subconscious. How many times have you have said to yourself, "I know I have a feeling about something," but you do not have all the information about it and therefore cannot predict the outcome in your conscious mind. This deep feeling or knowledge comes from deep inside you; it comes from the subconscious.

On the other hand, the conscious mind, which is an extension of the subconscious mind, was developed many millions of years after the primitive brain with a capacity to hold temporary information in its memory cells and process this information with its processing centers in logical ways to predict the future and control some body movements. But to do that, the conscious mind has to have memories and has to learn how to process them the right way, and that is the reason it takes the conscious mind time to develop, especially in humans: it needs to fill the huge capacity of its memory cells and to operate its processing centers in the right way. It needs to balance the knowledge of the subconscious mind or the collective memory as willpower, which takes time to develop, to gather more information and store it in the conscious mind's memory cells. This willpower is the power of the conscious mind to know what is good for it in the long run, since it can predict the future better than the subconscious mind, which wants to have things done right away without regard to future consequences.

So, willpower depends on the maturity and the knowledge of the conscious mind, but also on the intensity of the subconscious mind to extract its willpower on the conscious mind in the form of urges and feelings. In other words, you could be mature and very intelligent and still be controlled by your subconscious mind, because the urges and feelings that come from your subconscious are very strong. But since the conscious mind is a tool for the subconscious, the conscious mind tries to fulfill the desires and the urges of the subconscious, and the subconscious rewards the conscious mind with the feelings of pleasure and satisfaction.

For example, serial killers have a strong urge to kill that comes from their subconscious, and their conscious mind will try to implement this urge, but with cautiousness, by hiding their action. The subconscious will reward the conscious mind with feelings of pleasure and satisfaction once the urge is fulfilled.

In the next few pages I will mention a few urges and feelings that the subconscious mind imposes on the conscious mind, along with how the conscious mind tries to explain and fulfill them to the best of its knowledge in order to get piece of mind, pleasure, and physical satisfaction. The whole human body works better when the urges and feelings that come from the subconscious are fulfilled by the conscious mind.

The urge to preserve identity, which is the meaning of life, is very strong and very basic in all life forms including humans, and its intensity is different from person to person and from time to time. It is a basic urge because it goes all the way down to the genes themselves. Species do not mix genetically because their offspring would be totally different and they would lose their previous identity and become something else. There would be complications in the genetic codes of these new organisms, so they would not be able to function or reproduce, and would therefore be a waste of energy.

It is also basic in the genetic code to refuse to be taken by other species or even other groups within the same species. The human subconscious reflects that as an urge to stick to one's own kind, established through collective memory of past experiences of having less friction and spending less energy to adapt to the same group. The conscious mind translates these feelings and urges as attempts to find common denominators, such as size, color, look, educational background and geographical setting. A very strong common denominator is following the same leader; this leader could be alive, dead or imaginary. Since we are social animals we are designed by our genes to follow. When the conscious mind finds a common denominator through awareness of its surroundings, it transmits this information to the subconscious mind in the form of information molecules.

These information molecules create chemical reactions in the subconscious mind, forming new information molecules that are transmitted back to the conscious mind. The conscious mind perceives each of these information molecules as an urge or a feeling, depending on the molecular structure of the chemical message. It could be the feeling of belonging or love, or the desire to protect at all cost. Sometimes, when the urge to preserve identity is very strong, the conscious mind will justify these feelings by viewing itself (whether as an individual or as a group) as superior to other kinds of people that it views as inferior. This opens the door for racism. These feelings usually arise when the human becomes sexually viable, and can last for a lifetime because, after all, it is all about genes: you want your offspring to look like you. Many people would like to clone their genes so their offspring could be exactly like them. But because of the way genes work, it is much better to mix the genes within the same species but with far away groups in order to get the best benefits of the genes in the way of warding off disease or gaining better ability to adapt to different conditions.

This superiority urge or feeling that comes from the human subconscious gives you the "right" to kill and or consume other animals. This urge is the main force that twists the ability of the conscious mind to see clearly that the human race is part of nature's evolution, part of the animal kingdom in general, and part of the ape family in particular. Instead, the influenced human conscious mind sees itself as superior to other animals and as the product of a divine force. It sees humans as godlike creatures created in God's own image, and sees all other life forms as inferior to mankind. Everything is designed to serve humans and to circle around them, just like the Moon, the Sun and all the stars in our universe seem to circle our flat home planet.

The urge to deceive developed many millions of years after the development of the primitive brain, and it gave an advantage in survivability to the species that possessed it. These species were now able to deceive their prey with different techniques of ambush or by pretending to be something else in order to gain an advantage over their prey. They were also able to avoid becoming prey themselves through similar techniques of deception.

In humans, this urge to deceive that comes from the subconscious plays a major role in survival. It is usually implemented against other groups or individuals, especially in wartime when there is a need to defeat the enemy. Terrorists use deception by acting and looking like average people. Many businesses and advertising agencies use deception by highlighting the positive sides of their products without mentioning the downfalls. They try to gain as much energy (money) for the least amount of energy (money) invested. The deceiving urge is also the driving force behind con artists and just plan liars. They will justify these feelings and urges with their conscious minds by explaining to themselves that they are better than their victims or that they deserve what they can get. This urge could be very strong in some people, and will cause them to cheat and lie for no apparent reason.

The urge for suspicion arose at about the same time as the urge to deceive as a countermeasure. With so much deception going on, things are not what they appear to be, and if you want to survive you have to be suspicious about everything, especially of a strong and powerful enemy that has the power and the ability to hurt you. These genes that create the urge to suspect are embedded in the subconscious of all humans, but the intensity of suspicion is different from person to person and from time to time. The conscious mind will put some logical sense behind this urge by creating a conspiracy theory, and as everybody else tries to cheat or to gain something from them, humans will reason that the weak group of people are always in the right as under dogs against greater forces.

For example, after the events of September 11, which woke God up, when twisted, stupid terrorists who killed innocent people mentioned His name by using their subconscious instead of their conscious minds, God ordered me to write this book. There was a man in France who wrote a book accusing the USA of conspiracy in planning the attack on September 11 and destroying the twin towers. Now, this is an intelligent man, but the urge that comes from his subconscious to suspect overwhelms the judgment of his conscious mind against all facts and scientific evidence. His conscious mind explained this urge by reasoning that since the USA is the most powerful nation on Earth today, its government is looking for an excuse to go to war with other countries, and that this is the reason the US government destroyed the twin towers, killing three thousand of their own people. Thinking of the US as the most powerful country on Earth is the key element that triggers the suspicious urge of his subconscious. This urge is so powerful that his conscious mind put aside all the evidence and knowledge that the USA is a democratic country that is not running as a dictatorship, and that this country exports more knowledge and more "good" than any other country on Earth. This guy sold a lot of books because the urge to suspect is embedded in the subconscious of all humans, and only the conscious mind can tell the truth by using logic and its ability to see the future.

Telling the truth is also an urge that became embedded in the subconscious minds of humans later on in their evolution, when they lived in close-knit groups. It was a necessity to tell the truth and to be honest for the benefits and the normal operation of the group as a whole, for survival and bonding. The subconscious rewards the conscious mind with the feeling of pleasure whenever it tells the truth or the right information. The conscious mind explains this urge by knowing that dishonesty will be punished by the group or the group's leader, but that it is OK to lie to the enemy or to others that it perceives as an enemy.

Life and death: Life is recognized by the subconscious and collective memory because they have genes embedded in them as memories of how to survive, since they are products of life, but death is not recognized by the subconscious or by collective memory. The physical interaction and the experience of daily life are recorded by the genes and pass from one generation to the next in the form of urges and feelings in their subconscious with information on how to deal with experiences or events in the present based on similar past experiences. But in the event of death, the genes stop recording. There is no memory of death, and furthermore, dead bodies do not pass on genes.

The subconscious has the memory of pain, which comes from injuries or loss of limbs. It transmits these memories to the conscious mind as the feeling of fear in order to avoid similar conditions which could be encountered again. Because the event happened before and the subconscious was still alive, it recorded these memories and pass them on to the next generation.

But in the event of death there is no recording. So, because of the lack of those genes and of the information in the subconscious that records the death memories, there is no urge or feeling that you will die. On the contrary, the feeling that comes from the subconscious is that you will live forever, because it is a product of life, and the conscious mind gets this message as a deep knowledge coming from its subconscious. So the conscious mind gets the feeling that it will live forever. But the conscious mind is smart. It sees that death happens everywhere. It can put one and one together and see the future with its power of imagination, which comes from its long-term and short-term memory cells, and it comes to the conclusion that death will also happen to it.

But due to the lack of deep knowledge that would come from its subconscious that its death is certain, the conscious mind explains this paradox by understanding death as something that happens only to others and not to itself, or reasons that even if the body dies, the mind will live forever, maybe in the body of somebody else or in a place where you do not need a body, such as the Garden of Eden or any place it can imagine depending on the information it stores in the huge storage of temporary memories it gathers in its life time. After all, knowledge comes from memories. How you process them and who you are depends on what you know and the urges that come from the subconscious.

The urge to lead and to follow a leader exists in almost all animals that have a brain setup with both a subconscious and a conscious mind. It enables the weak or the unknowledgeable to follow the strong or the knowledgeable to safety or a food source, and in doing so preserves their identity by giving them a better chance for survival. Parents take care of their offspring, and when the offspring can take care of themselves they help to protect their own offspring, and by doing so preserve the species as a whole. So this urge to lead and to be led is embedded in the subconscious as a gene and becomes more complex in groups of animals that rely on their numbers to survive. So it is well developed in humans. All humans have complex genes imbedded in their subconscious that give them very strong urges to lead the group or to be led by someone. Throughout history humans always had leaders and followers.

The conscious mind tries to put a sense to this urge to lead the group, as it has more knowledge than the others in the group and better ability to lead the group to its destiny or to a better future, because it can see the future better than anybody else in the group. If an individual plays it right, in other words, satisfies the urges of the others in the group by showing strength and knowledge, he or she could be picked as a leader. Once they become a leader, they have the freedom to do anything they want within the group rules. And if they become the only leader (dictator) by knocking down their opponents (since other people in the group have the same urge to lead), it could unleash the dark side of their subconscious, since total freedom of control gives the subconscious total freedom to try to keep its leadership by using intimidation, fear, punishment and brutal force; total power is intoxicating.

The well-developed conscious mind of humans tries to put a sense or logic behind this urge to follow the leader, as the leader is more knowledgeable and stronger than the others, and can lead them to a better future. They can save energy by not trying to find the way themselves and by not questioning the leader. They need permission from the leader to do anything, and otherwise will be punished. They give the leader supernatural qualities, thinking that he/she can help them and take care of their needs.

But sometimes, reality hits them in the face. They can see that the leader is just one of them, that he has the same needs and is affected by the same conditions. He is nobody special; he is just an average or below average person. But the urge in humans to follow the leader is so strong, and imagination is also very powerful in humans, since their imagination is based on memories of the information they acquired through their lifetimes. So most people will see the leader in their imagination as supernatural; they will see the leader as God. This is because

God in their imagination is the only leader who has the power to control everything. And since they do not have the feeling that they will die, or in other words, since they feel they will live forever, they think that if they also give God this quality of being immortal, He will take care of them forever, even after their bodies die.

In their imagination, with the influence of the strong urge that comes from their subconscious to follow a leader, humans give their supernatural, illusionary leader the qualities of a human leader. The way they see it, He is smart, He knows everything, He is strong and He can help them when they need help. He can lead them to a better life on Earth, and if things do not go well on Earth, He can lead them to a better life after death, wherein all their wishes will come true, especially the wish to be disconnected from their bodies and from the material world. After all, in your imagination you can be anything you want to be, anywhere, anytime, and all you have to do is to follow Him because of the strong subconscious urge to follow the leader. He is the only leader and they are the followers, who pray to Him in various ways, just like a dog puts its tail between its legs when it approaches the leader of the pack.

Humans, who have a much larger brain, have to say how great the leader is no matter the conditions. Otherwise, He will get mad at them. They have to follow the rules He puts down on Earth through his various messengers, who are other humans with extreme urges and imaginations. Their imaginations use information and memories gathered throughout their lifetimes, with a twist of their own identities and conclusions. Those messengers act as leaders, leaders in thinking and bringing in new ideas, and play on the urges of the people to be led, since without followers there are no leaders.

The setup of the subconscious mind in humans assures this combination of a leader who will always have followers, no matter what his message is, or how crazy and unusual the ideas he brings or the promises he makes, especially about life after death. Since nobody came back to life after death, nobody can prove it is not possible, and since the leader is playing on the deep feeling of his fellow humans that they will live forever, they see him as a real leader. And once the conscious mind tells its subconscious that it has found a leader, either imaginary or flesh and blood, a person will believe it. This is because in the imagination there is no difference between real or imaginary. Whatever you believe is real, and it all comes from the mind. For example, if you order people under hypnosis to eat onions and tell them they are eating apples, they will enjoy eating the onions just as if they were eating apples.

The subconscious fills the conscious mind with feelings of love and compassion toward the leader. Even if the leader makes mistakes, the followers do not question him. In their minds they try to put sense and logic behind his mistakes, as though maybe they were being put to a test, or had themselves done something wrong and need to pull their act together and follow orders, or get punished now for a better future. The subconscious also puts a strong urge in the followers to protect the leader and what he stands for by giving him their energy or even their own lives, since the way they see it, without the leader they would lose their identity and have no future.

Therefore, the total strength of any religion or any leader, flesh and blood or imaginary, is really only the combined sum of his or her total number of followers, their total strength, knowledge and devotion. Each follower's strength and devotion comes from his or her genetic makeup and knowledge.

The urges and feelings that come from the subconscious are sometimes so strong that they can feel like a pressure. They can twist the ability of the conscious mind for logic and judgment to see the future, the truth and what is really good for you, since the conscious mind always tries to satisfy those feelings and urges that come from the subconscious in order to get the feelings of satisfaction, pleasure and joy that also come from the subconscious once the urges are fulfilled. But, the conscious mind can control, calm and lessen the urges and the feelings that come from its subconscious with knowledge and understanding.

The imaginary God perspective is different from person to person, since we are all a little different. We have different urges and their intensity is different. Our imaginations are different also, because imagination depends on memories and the ability to process them, which is different from person to person and could change with time within the same person. That is the reason why we see individual people changing beliefs or religions, or becoming more or less religious from time to time.

So, believing in God or in any other supernatural leader or creator has nothing to do with intelligence. You could be very intelligent and still believe in God, or you could have no intelligence and still believe in God, because the urge to follow the leader that comes from the genes in the subconscious is influencing us most of the time. But you do need some intelligence to understand where those beliefs come from and why those genes are there in the first place.

These genes that give you the urge to follow the leader act like gravity force which squeezes matter together. In this case, these genes bond the species or the group together for a better chance of survival, to give them the ability to follow orders from the leader. The leader is the common denominator for the group, and everybody circles around him. Without the leader, the group would have the feeling of being lost and lonely, and might disintegrate until they found the next leader. The urge to follow the leader is always present, as long as these genes are in the subconscious producing a substance that gives us the urge to follow the leader.

26. Life, Death and Life After Death

Our universe is made from a huge number of matter structures: smaller ones within larger ones. The smallest units of matter are the pairs of gravitons (the building blocks of all matter) that were created in the Big Bang from a different dimension of matter-energy. Those building blocks of matter grouped together in different numbers and shapes to form all the diverse subatomic particles. In turn, those diverse subatomic particles grouped together to form the smallest, lightest and simplest of all atoms: the hydrogen atom.

The fusing together of some of those hydrogen atoms through pressure and collision produced over a hundred heavier and more diverse atoms. The grouping of those atoms through gravity force, collision and chemical attraction created molecules. So, atoms and molecules are the building blocks of all matter structures in our universe. Anything smaller than an atom does not survive for more than a fraction of one second, unless it is all alone, since all subatomic particles always interact with their surroundings and change form. Therefore, all the structures you see in our universe are made from atoms and molecules: a structure within a structure. Humans are structures that are part of Earth's structure, which is part of our Solar System's structure, which is part of the Milky Way Galaxy's structure, which is a part of the structure of a cluster of galaxies, which are part of the structure of our universe.

All those structures obey our laws of nature, namely, gravity and energy. Our laws of nature put an order in the formation of all those structures. Therefore, there is a pattern, an order and intelligence, all on its own. All these matter structures in our universe have a lifespan. They are all alive, they are all in motion, and they all have their own time. Once those matter structures within the larger structure change their relative positions due to the ever present forces of gravity and energy, the larger structure changes form to become a different structure with its own new identity, and time restarts, giving birth to a new structure with its own lifespan. All matter structures are interchangeable and they change forms all the time.

The changing relative positions of matter structures also gives us the illusion of time ticking forward, since time is all about the relative position of matter structures. The present position is our present time, what used to be is the past and what will be is the future. The human structure, which is one of many life forms in our universe, is made from many smaller structures that are all interchangeable and changing with time.

A human is born, or comes into existence, with the help of the genes from his mom and dad. Under the right conditions these genes influence a chemical reaction that puts atoms and molecules in a certain order to create a new structure: a human being. The lifespan of humans can be up to one hundred and thirty years, but only if they inherited good genes from their parents, lived a simple life with no stress, no disease and no extremes, doing only the things designed for them by their genes (not too many people would like this kind of life), with plenty of good sleep, plenty of clean water, and the right food in moderation. Only one in ten million people will come close to this age of one hundred and thirty years. This is because all matter structures, including life forms, have a lifespan that depends on the number of interactions with outside forces.

The gene also has a lifespan depending on the number of times it can create new cells, and also on outside forces like radiation or molecules that can change the gene's molecular structure, altering the information carried by the gene of how to build the right cell. The human body contains over two billion bases and about thirty thousand genes that work in harmony and sequence to build the human body. Not all the genes in the human body have the same lifespan or are exposed to the same level of outside forces. For example, if the group of the genes that are responsible for building and maintaining the liver are overused as a result of too many interactions with the wrong substance or because of a disease that alters the gene structure, the liver will cease to function as an organ that acts as a filter in the human body. Once the liver stops functioning, the human body is doomed. Since all the human body parts work in harmony, without the liver the whole body will collapse and die.

In the future, genetic therapy will become more commonplace. It could be used to replace damaged or missing genes in the body and may extend a person's lifespan to beyond one hundred and thirty years. However, nobody will be able to live forever as the same person, because once you replace all the genes, even if this is feasible, you will become a different person. Nature does the same thing by letting you pass your genes to your offspring by mixing your genes with those of another person to start a new life by creating a new offspring. The offspring carries fifty percent of your genes and fifty percent of your partner's genes, but the way in which the genes fit together could make the offspring more or less like either partner. After a few generations of mixing genes, the genes will create a new person who hardly resembles the people of previous generations. It is the way of nature never to exactly copy the original forever, and to mix and match to some degree with every generation, since conditions and environment are constantly changing as a hallmark of time going forward. The mixing of genes allows for rapid development and selection of desirable traits, such as disease resistance or increased intelligence, keeping the species and its DNA going. The human body contains many parts or organs working in harmony together under the control of the brain. The primitive subconscious brain keeps our body temperature controlled, our breathing regulated and our hearts beating without us consciously thinking of these things.

The largest organ in the body is the skin, which covers and protects the body's internal organs, acts as a sensor for touch and external temperature, and regulates internal body temperature through sweat glands. Other organs that work together to allow the body to function are: the muscles for movement (every movement, even blinking, except some bowel movements caused by expending gas or gravity), the sex organs to pass on your genes, the legs to give you the ability to travel, the hands for work, the stomach to digest food, the kidney and liver as filters, the lungs to supply oxygen, the blood to supply the body with nutrients, water and oxygen and to take out waste products, and the heart to pump and circulate the blood. There are also the eyes, ears, nose and tongue, which are sensors that supply information to the brain via the nervous system. The conscious brain organ creates, stores, analyzes and compares information molecules to recognize objects, and/or to predict the future and to order the body to behave appropriately.

Many organs have multiple functions and all of them are made from a huge number of cells, usually multiple cell types that are unique to the organ they build. All those cells have a lifespan, and it is different from organ to organ. For example, bone cells live for up to a few months, whereas cells that line the stomach only live for up to a few days. Every cell in the body contains all the genes with all the instructions to make the whole body. Each new cell is created from the inside out, when supplied with the right nutrients through the blood. However, when the cells become more specialized in their functions during embryonic development, they only use the genes necessary to make a specific cell type, such as a brain cell, a skin cell or a heart cell. The genes in each of these cells have a lifespan and lose their ability to produce new cells for various reasons, such as loss of the integrity of a gene's molecular structure due to radiation. Also, each gene is programmed to only divide a certain number of times before it becomes senescent and dies. This is why as we get older our tissues do not repair themselves as well as when we are young, and we get wrinkles, heartburn, bad eyesight, and lose our minds.

Let's take the genes that produce hair cells for example. The body contains many different genes that produce different hair cells in different locations on the body, and they all have a different lifespan. Once some of those genes reach the end of their lifespans, they stop producing hair cells at that location, so the hair cells die and the hair won't grow anymore at that location. No shampoo or lotion can regrow hair at that location. Only gene therapy or the transplant of new hair cells that contain new genes can regrow hair.

All body parts and organs may be replaceable through transplant or gene therapy, either now or in the future. That is, every part except your brain may be replaceable, because your brain is who you are, the way you think, your memories, your imagination and your identity. If you replace the brain, or any part of it, you will become a different person with a different identity and different way of thinking. We see this happen all the time when the brain is damaged due to an accident, drugs, disease, genetic disorders, or from plain old age. In the case of old age, the genes that produce various brain cells reach the end of their lifespans and no longer produce brain cells the right way before they die completely and that part of the brain dies and thus ceases to function.

Some people may try to save their body and especially their brain (which contains their identity) after they die, by freezing it and hoping that science in the future will be able to bring them back to life. But the problem is that most cells in the body and especially in the brain contain about seventy percent water, which expands when it freezes and becomes crystal ice that punctures the wall or the membrane of the cell which contains and protects the DNA inside of it. Every cell in the human body contains DNA, and there are billions of cells in the brain. The cell protects the DNA and provides it with atoms and molecules to copy itself and make new cells. Without a functioning cell, the gene cannot function and the life cycle as we know it stops.

A dead brain may have a similar visual appearance to a living brain, but it cannot think, remember or coordinate vital functions necessary for life. All the billions of cells in the brain work in harmony as a result of their proper organization, which is encoded by the genes, with communication between the cells through atoms and molecules via blood vessels or electrons, via the nervous system. All this requires nutrients and energy gained by fusing carbohydrates with oxygen, which cells get from the blood supply. Now, if you remove any one of those three vital conditions (the correct organization of all those healthy brain cells, the ability of those cells to communicate, or the energy supply), that part of the brain won't function correctly or will die completely In the case of death, or when the brain is frozen, you lose all those three conditions. As the brain cells get damaged due to lack of energy, there is no way the brain can think, and the brain dies. And with it goes your identity and your feelings, and you do not care about anything. Your loved ones could be next to you suffering, and you won't do a thing, because you are back to nothing (where you came from) and the whole universe is nothing relative to you. This is true because we know the brain creates the outside universe inside of itself, and once the brain is dead, its universe dies with it. All the byproducts of the living human brain disappear when the brain dies. Its soul and its imagination disappear. It does not have feelings: no fear, no pain and no pleasure. It has total freedom; it has nothing.

Consciousness also dies and disappears. The perspective and awareness of everything in this universe disappears and dies like there is no universe at all. There is nothing. This universe is only for the living. You are out of the game of life since any life form is all about the relative positions of matter structures to other matter structures, which is also the exact definition of time. The present relative positions of matter structures gives us the present time, and as matter structures change their relative positions, time ticks forward. And as matter structures change their relative positions, they create new life forms.

So, after you die, the matter structures of which your body and your brain are made (molecules and atoms) will stay here and play their game of life. They will change their relative positions to create new and different life forms, new bodies and new information units by way of new structures. The game of life will continue and time will tick forward until the end of our universe, which will be in about fifteen billion years from now, when all the matter structures in our universe will be consumed by a black hole to become a different dimension and time will stop.

If you pay attention, you will notice that the number of years any person is alive has a limit; it is a finite number of years. The number of years before he was born is an infinite number, just like the number of years after he dies is an infinite number (as long as there is somebody that can keep count). All this means is that death and emptiness are the rule, and life is infinitely short and unusual. It is the temporary gift of the temporary relative positions of matter structures. These matter structures form a body with a lifespan that is infinitely short relative to the infinite universe time (The infinite universe that surrounds our infinitely small universe).

So, life is always short. The only difference is happiness, meaning the fulfillment of the desires and the urges of your body as designed by your genes (your blueprint). This also includes the stimulation of your brain without abusing or destroying other life forms while on the fine line of survival or self-defense to preserve your identity.

Life is like a mathematical equation: $0 = (^{[(1+2)_2 30]-80}) -5 = 0$. This equation means that if you dig deeply into the meaning of life, it will show zero, which is nothing, and then you could have an infinite number of lives and life forms in between those two zeros, in between those two nothings, just like a mathematical equation with infinite numbers of combinations and interactions, as long as you have a zero at both ends of the equation. It also obeys the law of time which says that if there is a beginning, there must be an end, and the end will resemble the beginning. So, if the beginning is zero, the end must be zero. It is just like an old man reaching the end of his life will look like a newborn baby (hairless, toothless and helpless, and with diapers) because he has gone through a full circle of life. He came from nothing and he is going back to nothing. He didn't exist before he was born and he won't exist after he dies. But he (whatever "he" is) did exist before in a different form, and "he" will exist in a different form after "he" dies.

Form is a certain number of matter structures in a certain relative position to each other to create a body, to create an information unit. Once the number of matter structures in the body changes, or their relative position to each other changes, the body form (shape) also changes and so does the information it represents, creating a new information unit and new knowledge through the way it interacts with its surroundings. This means that when something or somebody dies, it/she/he doesn't die per se, as if they had gone to a different unseen world or space like Heaven, Hell or a parallel universe. It just means that this person or thing doesn't exist anymore anywhere in the same form. They just changed form. The old form is not here or anywhere else anymore, and with the disappearance of the old form goes its old function, since any matter structure is also an information unit by the way it is structured and transmits knowledge in the way it interacts with its surroundings. Once the structure changes, so does the information it represents, creating different functions and different knowledge by the way it interacts with its surroundings.

So you can say that we die and are reborn all the time since we do change form all the time. Yes indeed, but we call it getting older. Did the five year old child you used to be a long time ago die? Not really. The five year old child you used to be just changed form to become you in the present time. You constantly change forms all the time. Therefore, what we perceive as death is really only changing forms.

We can't even agree on what stage throughout changing forms matter structures become a living human body. Does human life start before or after the sperm penetrates the egg, three weeks later, three months later, or nine months later? The true answer is only relative to your brain's knowledge and beliefs. But, the way we (God and I) see it, all life forms started in the Big Bang with the creation of matter to form unique matter structures, and death follows as those matter structures change form and the old structures disappear.

Therefore, we never really "die". We just change forms forever, for eternity, even going back to nothing, back to zero, and start all over again from nothing. There is no violation of the universal law of time which says that all matter structures have to have a full circle. If there is a beginning, there must be an end, and the end will resemble the beginning.

27. The Principle of Time Travel

As we learned in our previous discussions of time, time does not flow by itself. Time is always connected to the position of one matter structure relative to other matter structures. All matter structures in our universe are in constant motion due to the ever present forces of energy and gravity. It is through this motion that matter structures constantly change their relative positions, and this gives us the illusion of time ticking forward. The current position of matter structures gives us the present time, whereas the old position is the past, and what will be is the future. We know this because we have memory. We have a conscious part of the brain from which we can figure out the future position of matter structures. By remembering the old position of a matter structure and comparing that to its present position, we figure out the forces that moved those matter structures to their present positions, and use the same calculation to figure out the future positions of matter structures.

So, what happens if we reposition a certain number of matter structures to positions they had in the past, the way we remember them to be? It will give us the illusion that time went backward: the exact principle of time travel. To go backward in time we need energy to reposition matter structures as they were before, and we can only do this in a limited number of structures. We do not have all the matter-energy to reposition all the matter structures in our universe as they were before, since we are limited in speed and matter-energy. Our universe is always irreversibly losing matter-energy to black holes, meaning we can never retrieve that matter-energy. We are also losing energy in the form of photons, which we could never reach. Photons expand outward at the speed of light, and we could therefore never reach them and change their direction because our speed is limited by our universe's fabric. Therefore, our universe as a whole can only go forward in time. But we could reposition a limited number of matter structures here on Earth, and that will give us the illusion that we have gone back in time. The best example of this is to go to the masters of illusion themselves: to the moviemakers of Hollywood.

When they make a movie set in a previous era, let's say one hundred years ago, they use pictures and memories as a reference of what the world looked like back then. They can reconstruct a geographical setting, let's say a small town, to look the way it did one hundred years ago, by moving matter structures and repositioning them to the way they were one hundred years ago. They create the same houses, the same streets with gas lamps, horses and carriages, and maybe a few old cars that were operating back then. The actors dress the same way that people dressed one hundred years ago, and with the same kind of hair style. So, if you walk through this little town, you will have the illusion that you have gone back in time one hundred years, because all the matter structures in this town are in approximately the same relative positions they were one hundred years ago.

Another example of the manipulation of time is when people use plastic surgery to reposition their skin and hair to look younger, or in other words, to go back in time. Maybe in the future doctors will be able to reposition all the matter structures in the human body all the way down to molecular structures, and by doing that, the human being will not only look young, but will also feel young again. Nature has been doing this for billions of years since the appearance of the genes, by duplicating life forms to be more or less the same as they were before. A newborn son may look like his father did thirty years previously. So, to go back in time, all you have to do is reposition matter structures as they were before, to the way you remember them.

Time travel to the future is a different story. You do not have memories of the future, which means you do not know exactly what the future will look like (or more correctly, what the future relative positions of matter structures will be). However, in some cases you can figure out more or less what the future will be. You can move matter structures to whatever positions you imagine they will occupy in the future, and that will give you the illusion that you have jumped into the future.

To reach the future more quickly, you can slow or stop your body's movements and interactions. This means your body's own time will slow or stop relative to other matter structures that have normal motion. The easiest way to achieve this is by going to sleep, which slows the motion of your body. Your body ages more slowly in sleep because less of its matter structures are changing form through interaction with other matter structures. You do not eat or drink and you consume less air while you sleep, so your body changes its form more slowly, and that reduces aging. (Changing forms is another hallmark of time: the faster matter structures change form, the faster time ticks forward for them, and the more slowly they change form, the more slowly time ticks for them.) Furthermore, your conscious mind is shut off during deep sleep, and when you wake up eight hours later it is as if you have jumped eight hours into the future from the time you went to sleep. So sleep a lot: it is good for you, you will live longer and you will be farther into the future.

A similar effect is seen when the brain is in a coma. Perhaps in the future people will use hibernation or total freezing of the body to travel farther into the future.

There are other theoretical ways to slow your body's motions that are not really practical. One of these is to travel at a speed close to the speed of light. No matter structure can travel at the speed of light except for photons. At the speed of light matter structures lose all their spaces and become photons, which, as we have shown, do not experience time because there is no motion like there is in normal building blocks of matter. If matter structures move at the speed of light and change to a huge number of photons, they will lose their information because matter structures are also information units that transmit information by their unique multi-dimensional shapes, which give them their functions. Photons, on the other hand, are one-dimensional only.

For example, a spaceship is a spaceship because of its shape (shape is the relative position of all the matter structures of which the spaceship is made). Its shape gives it its function. At the speed of light the spaceship will become photons, and will not be a spaceship anymore. It will lose its shape, or in other words, its information, and will become a huge number of single points (photons) with no connections between them. But, in theory, the right spaceship can reach a speed approaching the speed of light. Let's say we can build this kind of a spaceship in the year 3535, which can accelerate to ninety percent of the speed of light using a huge amount of energy. Once it reaches this cruising speed, it does not need any more energy to cruise at that speed because there is no friction in Space, but it will need the same amount of energy to decelerate when it lands on Earth again.

We put one twenty-five year old person in this spaceship who has a wife of the same age and a five year old son. He leaves his wife and son at home here on Earth, and this young man goes alone on this spaceship at ninety percent of the speed of light for twenty-five years. During this time all the parts of the spaceship, including the lone traveler, increase in weight. The lone traveler notices that he needs more energy when he is moving around. The spaceship will also shrink to about half of its original size, with everything in it, including the lone traveler. Therefore, he does not realize that there is a change in size, since everything in the spaceship shrinks in the same proportion and the relative sizes remain the same.

So, at ninety percent of the speed of light space itself shrinks, and with it, so does time. The clock on the spaceship, which is about half of its original size by this point, also ticks at half its normal speed, but for the lone traveler the time is accurate since his heart rate stays the same at sixty heartbeats per minute. In other words, all the structures of which the spaceship is made slow their time by fifty percent, but the relative time among them stays the same because they are all in motion together at a speed of ninety percent of the speed of light. The lone traveler will also need a huge amount of energy to heat up his compartment, since the temperature of the spaceship when it moves at ninety percent of the speed of light will be only about three degrees above absolute zero.

When the clock on the spaceship shows that twenty-five years have passed by, the lone traveler knows that he is fifty years old and decides to return to Earth to celebrate his birthday. However, when he gets to Earth he finds out that he has been gone for fifty years (in Earth-time). His wife is seventy-five years old and his son is older than him, at age fifty-five, making for a beautiful and unusual birthday party.

We would get the same exact effect of space and time shrinkage if we built a house on the surface of a neutron star, lived in it for twenty-five years, and then returned to Earth. The strong gravity of the neutron star will compress the space of any matter structure touching its surface. The motion of the matter structure will become slower since it has less distance to go through. So the smaller the relative space, the slower the relative motion, the more slowly matter structures change positions, and therefore, the slower relative time becomes. And those are the facts of time.

So, if you want to slow your own time relative to other people, you have to compress your own space, and this can be done only by increasing speed (to close to the speed of light), or by placing yourself under a stronger gravity force (like on the surface of a neutron star), or by lowering the temperature to close to absolute zero, or by having less motion and interactions. These are the simple secrets of living longer or going farther in time toward the future, relative to other people.

As we said earlier, the only way to go back in time is to reposition matter structures to the same way as they were before, which will give you the illusion of going back in time. For example, let's say you can reconstruct your dead grandmother to her ten year old self by repositioning all the matter structures of which she was made, all the way down to the atomic and molecular structures of her body, with the exact relative positions as when she was ten years old. Now, if you succeed (an impossible task for now, but maybe not in the very distant future), you will have created your grandmother as she was when she was ten years old. It will be as though you have gone back in time or your grandmother has gone forward in time. She will look, feel and behave just like she did when she was ten years old a long time ago. Now, this little girl can grow up and marry somebody else other than your grandfather, since your grandfather is dead and you didn't recreate him or reconstruct her old surroundings, and she will therefore interact differently with her present surroundings. She will have her own different family, but it won't create the famous paradox of your whole family disappearing because you went backward in time and changed a factor in the past that could later on in time change the present conditions. In this case, you really didn't go back in time. The past time does not exist, since time is really an illusion, and it is the relative position of matter structures that gives you the illusion of time.

We hear many stories about time machines and especially about the use of wormholes. Even some scientists believe that in theory it is possible to use wormholes to travel forward and backward in time, but this is all nonsensical imagination going haywire, fueled by the urge of almost all human beings to believe in some kind of supernatural force, or even in UFOs. It is a symptom of the wishful thinking of mankind to have immortal life by building time machines. This strong wish or urge that comes from within the human mind can twist the facts. And the fact is that most people, including most scientists, do not know what time is all about, and that nobody except for God knows why time is really an illusion. (See the chapter "God and the Illusion of Time.") Furthermore, there are no wormholes, which supposedly resemble cylindrically shaped black holes that can stretch in size to many light-years. Black holes are round ball-shaped empty holes which the fabric of our universe tries to close by pushing its gravitons toward them with equal force from all directions to give them their round shape. The largest black holes are only a few miles in diameter.

But let's say we could create this cylindrically shaped black hole called a wormhole. It has one opening somewhere near Earth, and the other end of the wormhole is somewhere else in our galaxy, one light-year away. Any matter structure entering this wormhole will explode to become a different dimension and could be anywhere in the infinite universe in no time, and subject to different laws of nature. In other words, matter structures, be they human, clothing or tools, will lose their structure (their information) and becomes a different dimension. And therefore, we cannot control their destiny. But let's say we can enter this wormhole from one side and come out the other side one light-year away, in no time, completely intact. Now the fun starts! Using a powerful telescope (none exist) we could see ourselves on Earth: we could see what we did a year ago, as though we had gone back in time one year. We could also receive radio and television signals from Earth that indicate Earth-time as one year behind.

We remembered that we had a gas leak in our house which caused a fire eleven months ago, and since we went back in time one year, we can correct the problem and prevent the fire from happening in the first place. So, we enter the wormhole and come out the other end, where a spaceship takes us back to Earth just to find out that Earth-time is the same as when we left it, plus the few days we spent on our journey back and forth. So, what happened to our journey backward in time? The answer is: it was an illusion. The photons that carried the light, radio and television signals from Earth took one year to reach us because we were one light-year away from Earth. All this means is that because you travel faster than the speed of light, it does not mean you are going backward in time. It just means you could be at different points in space faster than the photons, and you have to wait for the photons to reach you and bring you only the images of the past, be they light, radio or television signals.

When we look at the night skies, we can see stars and galaxies far away from us. Some of them are billions of light-years away, but the truth of the matter is that we really do not see those heavenly bodies. We can see only the photons that came out of those objects in the sky and hit our eyes. These photons have traveled in Space for many years, up to billions of years, and are only information memories of the shape, size, motion and chemical makeup of those heavenly bodies, provided by the different wavelengths of the photons and the direction they came from. If we had the ability to be at any point in our universe in no time, we would see that those stars and galaxies are not there anymore. Some disappeared, some changed shape and form, they all changed their relative locations, and some new ones were created.

When we look at our Sun, the Sun is really not there; it is already eight minutes in our future, because it takes eight minutes for the Sun's light to reach us. All this means is that there is no real-time in our universe because it is in motion all the time.

So, going backward in time is impossible, unless you reposition matter structures to the same way as they were before, which will give you the real illusion of going backward in time. It is easier to go forward in time. All you have to do is stop your conscious mind from recording outside events, and wait for outside motions to reach the period of space-time in the future that you are looking for. Once you reach that point of time in the future, and restart your conscious mind, your present time will be the future you are looking for.

28. UFOs

We have all heard about UFOs. Some people believe they are real, and some do not. The word UFO immediately brings to mind a flying saucer piloted by alien beings that resemble us, with slim bodies and big heads. Of course they need to have big heads to store the large brains that helped them build a flying saucer that can travel through Space for distances of many millions of light-years to reach us and show themselves to a very lucky few people. The truth of the matter is that UFO stands for Unidentified Flying Object, and that means any unidentified object. Those are often manmade, which in most cases are objects like balloons, helicopters, airplanes, secret prototypes of airplanes or flying spy machines.

The phenomenon of UFOs could also include naturally occurring objects like those that fall from Space, changing direction as they hit the atmosphere, or balls of static electricity that can move very rapidly from one location to the next and then disappear. There are also mirages in which you can see objects that are not really there, which happens as the atmosphere reflects light coming from Earth's surface and those lights move, change shape or disappear altogether as the atmosphere moves and changes its shape. Even a car traveling on a hot day on a flat desert road will seem like it is floating in the air.

There are also many hoaxes and lies involving sightings of UFOs made by people who do these things for different reasons. Some of these reasons have included cover-up for the building of secret flying machines, while others are about fame or money, and some people even do it as a joke. All of these hoaxers are targeting the natural urge (the need) in people to believe in some kind of supernatural phenomenon or conspiracy. There are people who really believe in UFOs, and some people have even claimed that they encountered beings from outer space. Those people have a strong imagination that can twist the facts. Their imagination goes haywire as it is influenced by strong urges and feelings coming from their subconscious, such as fear or a wish to believe in something that is not there or didn't happen. The facts can also be twisted by drugs, fatigue or strong magnetic fields that can alter the normal brain functions so that a person imagines or sees things differently from what they are.

The fact is that we never have been, and never will be, visited by intelligent beings from outer space (unless you call an asteroid an intelligent being) for six reasons: first and foremost, as we discussed earlier, the chance that there is an intelligent civilization in our universe is very slim to none.

Second: the chance that there is an intelligent civilization in our universe with technology and knowledge that is advanced enough to build a spacecraft that can travel at fifty percent of the speed of light is very slim to none.

Third: even if there is an intelligent civilization somewhere in our universe with such knowledge and technology to build this kind of spacecraft, they should know it is not possible to build a spacecraft that can travel faster than the speed of light or even approaching the speed of light. It is also not possible to build a time machine. Another intelligent life form will have the same laws of physics and should know these laws if they are so advanced. Therefore, they should know that they cannot reach us even if they are looking for other intelligent beings.

Fourth: the great distances between different points in our universe are prohibitive for space travel. These distances can cover many millions and billions of light-years, and if the spacecraft has to travel for many thousands, millions or billions of years at fifty percent of the speed of light, it is very difficult, because at that speed even a grain of sand can cause massive damage by collision with the spacecraft. Also, it is difficult to imagine that anybody could afford or survive such a long journey.

Fifth: even if there is an advanced intelligent civilization nearby in our own galaxy with sufficient knowledge and technology to build a spacecraft that can travel at half the speed of light, it would still take them many years, if not thousands of years, to reach us. They would need twice the time relative to photons (which always travel at the speed of light) to reach us, so they would be wiser to reach us through radio and TV signals. This would be much more cost-effective, use much less energy, and be much faster than reaching us with a spacecraft. And they should have this technology if they are so advanced.

Sixth: even if those advanced beings from different planets reach us with their amazing spacecraft, why would they go into hiding in the bushes and show themselves to only a few people? What have they to fear? With such advanced technology, they should be able to make contact with our presidents Abraham Lincoln or George W. Bush.

So forget about Hanger 51, which supposedly contains the remains of an alien spacecraft along with the bodies of beings from outer space. It is not true. I wish it were. It could have given us an inside look at different life forms and new technologies. I do not think anybody could keep such a big secret, because this knowledge would give mankind a giant leap forward.

29. Randomness and Uncertainty

Randomness and uncertainty: not knowing the outcome and not knowing the future relative position of matter structures. Therefore, randomness and uncertainty are simply a lack of knowledge, just like darkness is a lack of light. Everything that happens in our universe has a reason and is the result of a chain reaction of matter structures as they engage other matter structures to create different matter structures and/or change their relative positions as they follow our universe's laws, which are mainly the laws of gravity and energy, and which always involve motion.

Our universe is made from a huge amount of matter structures: smaller ones within larger ones. Most matter structures are either too far away or are hidden behind other matter structures, and some are too small for us to take any notice. Every matter structure is also an information unit by way of its structure, and creates knowledge by the way it interacts with its surroundings. Therefore, there is a huge amount of information in our universe, which our brain cannot absorb, either because the brain is not aware of it or because it is just too much information for the brain to store or process. Nonetheless, all this missing information that we either cannot see or cannot process is a factor in creating the physical future. Therefore, randomness and uncertainty overshadow knowledge by many thousand fold.

The brain is like a time machine: it is able to predict the future as it creates the illusion of time. But to predict the future correctly, the brain needs the right information and knowledge of how matter structures move and interact with their surroundings. Therefore, the absence of information in the brain about all the matter structures involved will result in lack of knowledge and the inability of the brain to predict the right future positions of those matter structures. The future positions of those matter structures will be uncertain: a random possibility within all the possible choices.

For example, I can flip a coin up into the air in a closed empty room, and the coin will flip in the air and land on the flat floor. There are only two possibilities for which side of the coin will face me once it lands motionlessly on the floor: either heads or tails, and nothing else. A monkey will not appear from nowhere and replace the coin. But which side of the coin will face me is uncertain; it will be a random possibility, one of two possibilities. So if I want to eliminate the randomness, that is, to know the future position of the coin before I flip it in the air, I have to have knowledge about the starting relative position of the coin, the amount of energy it receives, and the way the coin interacts with air molecules, gravity force and the floor. Once I have this knowledge, I can predict the future position of the coin as it will lay on the floor by eliminating the randomness, or in other words, by eliminating the uncertainty. I can also flip the coin once, see how it lands, and repeat the same process of flipping the coin under the exact same conditions. Then the coin will land on the same side as before.

But the same exact conditions are hard to come by, since there are always fluctuations in the forces of gravity and energy due to the motion of matter structures as time ticks forward. The motion of Earth around the Sun as Earth changes its relative position and distance from the Sun, the motion of the Moon around Earth, or even an airplane flying above, will cause fluctuations in gravity force. Temperature changes will also cause changes in the density of matter structures, which will therefore interact differently among themselves. So, total knowledge is a hard thing to come by, and with this lack of knowledge comes the inability of the brain to predict the future accurately.

Accidents are caused by a lack of knowledge. If you know that an accident is about to happen, you will avoid it or stop the chain reaction that can lead to an accident in the first place.

Therefore, randomness and uncertainty are part of life. You will never have all the knowledge all the time to predict the future correctly, or even to create the future you are looking for.

30. Knowledge

Knowledge. It sounds like a simple word. But what is it? How do you know what you know? What lets you know what you know? How does knowledge manifest itself physically? We always associate knowledge with the human brain, but is it true that knowledge belongs to a human brain only?

I always wondered how a photon that is made from a single block of matter and does not have a brain knows that it should always travel at the speed of light. If I light a match, I create new photons, and they already know at what speed to travel. How do they know that? If I throw a stone straight up into the air, it goes up to a certain height and then falls straight back to Earth. But how does it know to perform (do) all those functions and movements when it does not have a brain? A leaf falls off a tree, it twists in the air and gently hits the ground. How does it know to do all those things? How does a virus with no brain have the knowledge to identify and penetrate only certain living cells? How do the genes know how to build a body? How does the human brain create and hold knowledge? How do medical and law students create and possess the knowledge to become doctors and lawyers?

So, for knowledge about knowledge, I again asked the God of Knowledge for advice, and He answered me right away. What God told me was very interesting. God said: "Knowledge does not just appear out of thin air; knowledge is always connected to the unique shape of matter structures and the way they interact with their surroundings, which are other matter structures, and also to the fabric of your universe. Therefore, knowledge is the end result of interaction, and as you see, knowledge always involves the motion of matter structures, and time ticks forward." True.

The fabric of our universe is made from huge numbers of single gravitons that are repelled from each other by lines of negative energy. With their numbers and their motion, these single gravitons impose speed limit and gravity force on the building blocks of matter in our universe.

The photon does not "know" at what speed to travel; its unique shape determines its speed as it interacts with the fabric of our universe. The photon is made of two gravitons with no distance between them. Therefore, it is the unique shape of the photon which creates an information unit and dictates the way it interacts with its surroundings, and this is knowledge.

When I throw a stone straight up in the air, I give the stone energy to counter the gravity force Earth is creating. But I give the stone a limited amount of energy, to go up to a certain height only. In other words, it is enough energy to counter only a certain number of single gravitons. These gravitons, which make up our universe's fabric and carry gravity force, are falling toward Earth's surface with a gravity force of one "G", and therefore these gravitons that are coming towards Earth's surface will constantly deduct some of the stone's energy. Once the energy the stone is carrying by way of its motion gets to be zero, the stone will stop its upward motion and will fall back to the Earth's surface as it gets carried toward Earth by the motion of the gravitons. So the stone does not have the knowledge of what to do; it just follows our universe's orders as its energy interacts with the fabric of our universe, and its unique shape and density interact with air molecules.

The same goes for the leaf. Its unique shape will interact differently with its surroundings to create different motions, as it has different surface areas and densities that interact with air molecules and gravity force. Therefore, its unique shape (its information unit) will interact differently with its surroundings, to create different motions and different knowledge of what to do.

The virus has no brain and no real knowledge of what to do when it penetrates only certain living cells. The knowledge of the virus to recognize and penetrate only certain living cells comes from its unique surface molecules, which recognize only certain molecules at the surface of living cells and penetrate them through chemical reactions. Once the virus is inside the living cell, it uses the cell's genetic mechanism to copy itself and to make new viruses, all through chemical reactions. So, the virus's knowledge comes from the uniquely shaped molecules at its surface that act like a key to open only certain locks, locks that the living cells create by having their own uniquely shaped molecules at their surfaces.

The gene creates the body without any real knowledge. The genes create the body only by way of their presence, as their unique molecular structure, which makes the genetic code, creates other uniquely shaped molecules (proteins) in a series of chemical reactions. The human body is created by about thirty thousand different genes. Each gene has a unique molecular structure and will therefore interact differently with its surroundings to create a unique substance. The physical connection of all these different substances is what leads to the creation and organization of the body as it is. Most genes will produce their substance only under very precise conditions, in a certain tissue for example, or in the presence of a substance that is made by certain genes. All this happens by way of chemical reactions in which proteins with unique shapes interact only with certain other molecules to produce a new molecule, a new substance. Therefore, there is control: there is sequence, timing, order and knowledge, just by way of the shape of molecular structures and the way they interact among themselves to create the living body.

The human brain contains knowledge using the same natural laws as everything else in our universe: by way of the unique shapes of matter structures and the way they interact with their surroundings to create different matter structures, different information units, and therefore, different knowledge. The brain creates knowledge by fusing copies of memory molecules (because of which memory stays intact) among themselves to create new molecules (new information units), and the reading of these new molecules through chemical reactions in the brain is the new knowledge. And as you see, knowledge always involves the motion and physical interaction of molecules, and time ticks forward. The copies of memory molecules can come from short-term or long-term memory cells, or from collective memory cells, and therefore, the knowledge the brain possess is limited to its stored information units and the way these information units interact with their surroundings, which is based on the brain structure as a whole.

Knowledge does not come out of thin air. You either have the knowledge that manifests itself physically in the form of the unique shapes of matter structures, or you do not. There are no miracles. Everything has a reason, is the result of a chain reaction, and follows the laws of our universe.

The process of creating correct knowledge is the same for creating incorrect knowledge, since there is no right or wrong and good or bad in nature. It is only relative to your own knowledge. Therefore, the process of creating knowledge for a genius is the same as for an idiot; it always involves the interaction of molecules with their surroundings inside the brain.

But the way nature intended, the right knowledge in the brain is purely for the survival of the individual, to preserve its structure and its identity longer through time, which is the meaning of life. Without the right knowledge to survive, the brain, along with the body it controls, will hit a dead end.

The brain has the basic knowledge to survive in the form of right or wrong / good or bad feelings that come from its subconscious in the form of molecular structures of a certain shape that are produced by molecular gene structures of a certain shape. From birth, the brain learns what is right and wrong for its own survival through trial and error and the experience of pain and pleasure. So it avoids the pain and seeks pleasure, as it is designed for that purpose by the genes.

The brain is just a tool, a part of the body that gathers information via the body's sensors, and recreates the outside universe inside of itself. The brain creates the illusion of time, and with it can predict the future. But to predict the right future, the brain has to absorb, read and become synchronized with the right information, with the new knowledge the outside universe is creating by way of its motions.

The genes that build your brain give you the capacity to learn, to create and to produce knowledge, and it's up to the environment, to the surroundings, to fill that capacity. A dog is a dog because his genes built him that way, and even if he spends all his life in a university, he won't get smart like a human being would, because the human's genes build their brains with much more capacity to learn. Even among the human species the ability to learn and understand differs from person to person, since DNA is different from person to person (except in identical twins). Therefore, it's possible that there are many people in Africa or in India, for example, who have high brain capacity, even genius level, and yet their brain capacity doesn't fill up because of their surroundings and conditions. Being in the wrong place at the present time (lack of schools and education) won't allow their brain capacity to fill up and excel in the right fields for which their brains were designed.

Medical and law students become doctors and lawyers only after they learn their trade in school. They create memory molecules out of information they receive in school. The information comes into the brain via the body's sensors, mostly from the ears and eyes in the form of electron waves, reshaping available molecules inside the brain. These molecules become memory molecules whose uniquely shaped molecular structures represent information units. These memory molecules are stored in short-term or long term memory cells. These memory cells can create copies of memory molecules they possess on demand, from different departments of the brain via electron signals. The fusing together of some of these copies of memory molecules creates new molecules, new information units, new knowledge, and the reading of these new information units in the brain, as the brain acts like an electronic scanning device, is the new knowledge, thought and feeling. So, without these memory molecules which capture past experiences, events and information by way of their unique structures, and without the ability of the brain to process and read them, these medical and law students wouldn't be able to become doctors and lawyers.

To become doctors and lawyers these students have to pass tests of their knowledge, of what they learned in school. Their individual test scores won't be the same, just as their knowledge won't be the same, since all the students are a little different from one another. Brain structure is different from person to person, and with it, the ability of the brain to receive information and process it. The students are also exposed to different conditions such as drinking, sickness and lack of sleep, all of which affect the brain's ability to absorb, process and produce information. So, the knowledge that can be produced is different from person to person and is based on information which manifests itself as memory molecules and the way the brain processes these molecules.

The human brain contains huge amounts of information in the form of memory molecules, as it constantly gathers information throughout its lifetime and makes memory molecules out of it. The human's brain also possesses information from birth in its collective memory in the form of genetic molecular structures, which are basically information units and act just like memory cells. The huge amount of knowledge and its complexity in the human brain is a byproduct of a huge amount of small matter structures of different shapes moving and interacting among themselves in complex ways.

All the knowledge the human brain possesses is the reflection of the unique shapes of matter structures and their relative positions in the brain as they move and change positions and formations. Without those matter structures in the brain, or without the ability to read them, there is no knowledge.

As you see, knowledge itself is not real; it is only an illusion created by the brain. What we call knowledge is only the reflection of the shapes and relative positions of matter structures and the way they interact with their surroundings and to the fabric of our universe as they follow our universe's laws and possibilities.

Therefore, all matter structures (including our brain) contain information by the way they are structured and create knowledge by the way they interact with other structures.

So, is knowledge real or not? Let me know.

31. The Meaning of Life

In the search for the meaning of life, I asked a couple of dogs eating from a trash can, "What is the meaning of life?" The dogs looked at me for a few seconds, wagging their tails, and went back to continue their meal. After they finished their meal, the male dog started to hump the female dog. I asked the same question of a stone that was standing out from other stones by its shape and color, and I didn't get an answer. It was completely silent. It didn't make a move. So in my desperation I asked God and I got the answer only after nine months (He was on a business trip, exploding and creating new universes). His answer was: "You are a product of nature. Nature has the answer." God is right.

For nature, the meaning of life is the struggle to preserve identity (structure) for as long as possible using the matter-energy created in the Big Bang. This matter-energy that was created in the Big Bang is limited in its quantity and its lifespan, and is the reason for the struggle to preserve identity. Everything in our universe needs energy to preserve its identity, its structure, to balance the ever present force of gravity. You will always need greater energy than a structure has to change it and to therefore change its identity. Everything in our universe tries to preserve its identity (its structure) by consuming other identities (other structures).

Identity, which is also an information unit, is always a reflection of the shape and relative positions of matter structures. All the matter structures we see and feel are made from a certain number of molecules and atoms. The relative positions of those different atoms to each other in a structure creates the structure's shape, its function, its information unit, and therefore, its identity. The identity could be any matter structure. It could be a simple thing like an electron, which resists being swallowed by its much larger atomic nucleus by spinning very fast. Only energy stronger than the energy of this electron can make it collapse to its nucleus and lose its structure, its identity, to become something else. This happens in neutron stars, where gravity force is so great that the electron loses out to the much stronger gravity force and falls into its nucleus to become part of a neutron star. It loses its identity (its structure) as an electron to become part of the neutron star. The neutron star also spins very fast to avoid becoming a black hole, in which case it would be of a different dimension and would lose its identity.

A virus always tries to exploit the replication machinery of cellular DNA to make other viruses just like itself in order to preserve its identity (its structure). A lion kills and consumes other animals, which are different structures, different identities, to preserve his own identity. Without consuming other animals, which gives the lion matter-energy, it will die and its unique structure, its identity, will disappear. A cow consumes grass and grain, which gives it the matter-energy it needs to preserve its identity. So, by consuming other life forms, animals can preserve their identities. Because of this struggle to preserve identity, only the strong survive, and life forms develop various ways to preserve identity. This is the driving force behind the evolution of life on Earth. Some animals became bigger and stronger, some became smaller and faster, some had the ability to adapt to different conditions, and some used their huge numbers to preserve their identities by replicating themselves faster than others (such as mice).

The maintenance of identity over many generations is ensured by the replication of genes and is the meaning of life for all life forms. Since they all have a limited lifespan, the only way to preserve identity for longer than that lifespan is through reproduction, and so they pass their own DNA to the next generation. Once reproduction is over, most animals (such as flies) die shortly afterward, as they have served their purpose and must make room for the next generation. Some creatures even offer the matter-energy of their bodies to their offspring in various ways (such as breast feeding or, more extremely, like certain male spiders who are cannibalized by the female after mating).

The meaning of life as embedded within our genes manifests itself in the human subconscious as an urge to reproduce, to have offspring, to preserve its own structure, its own identity. Humans will go to great lengths just to satisfy this urge, and it is noticeable especially when there are obstacles in the way of reproduction (such as the inability to get pregnant). Humans will try to replicate themselves with other humans who share their looks and qualities or with other humans who have qualities they like and/or with which they identify, in order to share their life or to produce offspring that will resemble their own identity and will preserve their own identity in the future. This strong urge to preserve identity is the main cause for racism and gives the feeling of superiority. It is the urge not to mix with different individuals or groups because you are superior to them.

Everything humans do is for the sake of preserving identity: the way they look, the way they act and the clothes they wear. A policeman tries to show his identity by the uniform he wears. The judge puts on the judge's robe in a courthouse to show that he is the judge. Humans construct buildings and monuments with unique structures to preserve identity through their lifetimes, all the way to the cemeteries, which have unique curb stones, letterings and names. All wars and human conflicts arise from the need to preserve identity or from clashes of identity. Individuals and groups of humans invade neighboring and distant territories to gain energy in the form of resources, slaves and land for their own expansion and to preserve their own identities. Furthermore, the invading group often imposes its own ideas of political and religious belief (which represent identity) on other groups by force in order to convert them or kill them, and to therefore destroy the opposing identity of this other group of humans. Humans will go to great lengths, even sacrificing their lives, to protect their own identity or beliefs.

Terrorists destroyed the twin towers on September 11 because the twin towers represented the symbol of free world trade, which was in contrast with their own identity, lifestyle and beliefs. They saw the free world as a danger to their own identity. They knew (after all, they had a little brain) that they couldn't destroy the USA, so they destroyed the identity that the twin towers represented and also tried to destroy the Pentagon, another symbol of US strength that also helps the US to protect and enforce its identity.

Like other matter structures, identity is four dimensional. Its unique shape as relative to other structures represents information, a message, and therefore an identity, and it needs energy to sustain its shape. Since it is not alone, there are always other forces that try to take its energy and by doing so change its structure, or in other words, its identity. A house of worship is a matter structure with a certain shape that sets it apart from other houses of worship. It has its own unique structure inside and out in order to represent an idea, an identity to which certain people can relate. This structure, like any other structure, needs energy to maintain its unique shape, since there are other forces, such as gravity and weather, that always try to change its structure, its shape, and if a force is greater than the energy the structure holds, that structure will be destroyed and its identity will vanish or change form, as can happen with an earthquake or a bomb, and only energy can put it back together.

Thinking, imagination and ideas are the reflection (the reading) of certain molecular structures in the brain, and as the structure of a molecule changes, the ways of thinking, imagination and ideas also change. The conscious mind tries to keep the molecular structure that gives it the feeling of pleasure by satisfying the urges that come from its subconscious. An urge is created by a molecular structure, and the reflection of this molecular chain reaction by the conscious mind will create an urge or a deep feeling that you are doing the right thing. It is like following a leader with whom you identify, or belonging to a group, any group with which you identify. It could be the Ku Klux Klan, Green Peace, a church group or a Muslim terrorist group.

Humans try to transmit information with the way they look, which is also a structure and is therefore an identity. Religious people put on special clothing, cover their heads or grow beards, and try to preserve this structure as long as they can, or as long as they identify with it. All the different flags and anthems of different countries symbolize different identities, different structures, which the governments and often the people of those different countries try to preserve.

In general, as people age they try to preserve the look (the identity) of their fertile years. But what happens when a person has a disease like cancer? Would he or she try to maintain their new structure? The answer is: no. Since all structures are made from smaller structures and an identity that maintains their structure with energy, there is always a struggle to maintain the identity by drawing energy from surrounding structures. Cancer is no different. Cancer tries to draw energy from the body to sustain its own structure, while the body tries to sustain its own structure by fighting back, and the strongest wins. Whichever entity wins leaves its mark in life longer, and that is the meaning of life: to sustain a structure, and therefore, an identity, as long as possible.

But all structures do change in time. It is the law of our universe. The play between energy and gravity causes movement in matter, causes time to tick, and creates the lifespan of all matter structures. This also applies to identity, which is under constant pressure to change and evolve. The change in structures is the evolution of structures, but that is another story.

The dogs and the stone I mentioned at the beginning of this chapter actually helped me to understand what the meaning of life is all about. When I asked the dogs what the meaning of life is, they answered me with their actions. They were eating, consuming energy to preserve their structures, their identities, in the present time. And they tried to pass on their genes to preserve their structures, their identities, into the future.

The stone also answered me by not making a move (since it is a stone). It told me with its silence that it is structured like a stone, and that to change this structure, to make something else out of it, to change its form and to therefore change its identity, I do need energy.

32. God and the Illusion of Time

One day, while taking a shower, I was thinking about the riddle of time. What is time, and why could it only be an illusion? (Illusion: something that is not real, is not really there. It is only created in the brain, and so is what we perceive to be real). This question has been on the minds of many human beings throughout history.

For many thousands of years, scores of philosophers, scientists and laymen have tried to answer this question without success. This question intrigued me the most, because once we know what time really is, we will know the secrets of this universe and will know God much better. So I called upon God, since He is the only force in our universe that knows all the answers, and to my surprise He answered me right away. (I guess he liked me clean.) The answer was: "Only the present time is real and it is motionless. The past is an illusion, just like the future." God is right.

We know that time is always connected to the relative positions of matter structures. The current relative positions give us the present time. However, the relative positions change because of the underlying ever present forces of energy and gravity. These two forces create motion, which changes the relative positions of matter structures, and we get the illusion of time ticking forward or backward.

We also know that the relative positions of matter structures to one another other create a body, and as those matter structures change their positions, again due to the interplay between the forces of energy and gravity, the body changes form and we get the illusion that this body of matter is getting older or younger.

The relative positions of matter structures also give us an information unit by way of their unique shapes (as long as somebody can read it). For example, letters represent information units because of their shape, just as a door lock will open only to a specifically shaped key that represents an information unit by way of the relative positions of its matter structures.

So, there is a direct correlation between time, information and a living body: they all have a presence by way of the relative positions of matter structures to each other.

For example, the position of the Sun relative to Earth gives us the time of day. When the Sun is above you at ninety degrees, you know it is noontime as long as the Sun remains directly above you. As the Earth spins around itself, the Sun's position changes relative to our location on Earth, and we get the illusion of time going forward. After about twenty-four hours, the Sun's relative position is about the same as it was about twenty-four hours ago, and we get the illusion of time going backwards; it is noontime again.

The relative positions of the huge number of matter structures of which the Sun is made also creates the Sun's body. But those huge numbers of matter structures always change their relative positions due to the interplay between the forces of gravity and energy, and we get the illusion of the Sun getting older. The Sun loses more matter-energy through the emission of photons and solar wind than it receives through the collision of matter structures crashing towards its enormous gravitational field. There is a constant loss and gain of matter-energy and the matter structures in the Sun change their relative positions all the time, which makes the Sun a living body, and as with any living body, it has a lifespan. It will change forms, and due to the interplay of the forces of gravity and energy it will change again to become a red giant, and then in a few million more years, a white dwarf.

The relative positions of the matter structures of which the Sun is made also give us its information unit. We know it is the Sun because of its shape, or in other words, its structure. Once the relative positions of the structures in the Sun change, again due to the forces of gravity and energy, it can become a totally different information unit, like a red giant or a white dwarf. So, the changing relative position of matter structures gives us the illusion of time flowing either backwards or forwards.

So, where does the illusion of time come from? It all has to do with the motion of matter structures. As matter structures move due to the interplay of the forces of energy and gravity, they change their positions relative to other matter structures around them. Therefore, their old positions become the past, their current positions create the present time, and the future positions of matter structures is the future itself. But how do you know that these matter structures are in motion or have changed position when your sensors (your nose, ears and eyes) only give you bits of information from the real, or present, time? Just like God said, "Only the present time is real and it is motionless." In other words, matter structures can only be in one relative position at one time, and that time is the present time, and therefore, the information your brain receives from its body's sensors comes only from the present. It does not come from the past, because the past is not here anymore, and it does not come from the future because the future is not here yet, and therefore, the sensors cannot detect motion or the change in the relative positions of matter structures.

The sensors can detect only the present relative positions of matter structures by taking a snapshot of the relative positions of matter structures in the present time only. But the present time itself changes because of the motion of matter structures. So, motion is an illusion of the brain. So, how is the illusion of motion created in the brain? It all has to do with the imaginary power of the brain. The imaginary power of the brain helps you to cope with reality (the motion of matter structures outside of the brain) by activating the memory cells in the brain that record events (the relative positions of matter structures) in real-time (the present time) as the brain receives and stores the information coming from its body's sensors. The brain stores the information in its memory cells, and whenever the imagination needs these memories, it pulls them out of the memory cells, and the past time, or the past relative positions of the matter structures involved, comes back to life.

The brain creates the outside universe inside of itself through the interactions of chemical reactions and electrons. The brain does not see. It therefore recreates. The brain reconstructs perceived motions of outside matter structures within itself, as an illusion. The illusion of motion inside the brain is usually synchronized with outside motion, but one twenty-fifth of a second behind. This is about the amount of time it takes to translate the information the brain receives from its body's sensors to chemical information. Therefore, the brain cannot create (it cannot see) outside motion if the motion of outside matter structures is faster than one twenty-fifth of a second relative to other matter structures around them. (Magicians and illusionists know the hand is faster than the eye.)

So, the brain can detect motion only by recording the present positions of matter structures relative to other matter structures and by storing these recordings in its memory cells. Once these matter structures have moved, they change their relative positions and the brain can compare the present positions of these matter structures with the old positions, as recorded previously in its memory cells, and it knows that there is a motion or that there was a motion because the relative positions have changed. (The change in the relative positions of matter structures is always due to motion, which is the fourth dimension.) The byproduct of this knowledge is the **ILLUSION OF TIME**.

Therefore, if there is no motion, or if the relative position is unchanged, time does not flow from the past to the present and then to the future. The brain can predict the future only by reading and comparing memories from the past, or the past memories to present information. So, present information entering the brain becomes the past almost immediately, or more correctly, in about one twenty-fifth of a second, since it takes the brain about one twenty-fifth of one second to record information and turn it into molecular information. Therefore, these new information molecules act and behave just like regular memory molecules. The only difference is the sequence. By reading these memory molecules in sequence, the brain can put an order and recreate the outside motion in speed and direction, following the pattern in order to predict where the matter structures involved will be in the future.

For example, matter structures are put together by the genes to create a lion's body that by the shape of these matter structures creates an information unit that is unique, one of a kind, and it represents a lion. This lion is charging or moving towards a deer that by itself is another matter structure, another information unit by way of its shape. (Shape is more than skin deep. It goes all the way down to the relative positions of all the matter structures of which a body is made. This makes each body unique and one of a kind.) The deer detects this information unit, this lion, with its vision, and recognizes it as a foe, as a predator. He also detects the motion of the lion towards him, as the deer's brain constantly records the changing positions of the lion relative to its surroundings, and relative to the deer in its memory, and almost instantaneously, compares these memories to the present position of the lion. Therefore, the deer creates the illusion of motion in his brain and has the illusion of time, and with the illusion of time, he can predict the future. He will become the lion's next meal!

To avoid this future, he runs away to survive, to preserve his identity (his information unit) and to survive for another day in the jungle of life. This is the meaning of life: to preserve your structure, your information unit, for a longer period of time by applying energy.

So, the past is not real. It is not there anymore. It has changed form to become our present. The past is only imaginary; it is only in your mind. It is based on memory cells and the power of your imagination to recreate it. These memories can be twisted or degraded depending on the quality of the memory cells that are subject to various conditions, the altered connections among them, and the ability of the imagination to process and read all this information in the correct sequence.

The imagination does not have to be visual like a dream. It can manifest itself as a feeling of you doing the right thing, or as a real illusion. As information comes in real-time via the body's sensors the imagination can compare past positions of matter structures that are stored in the memory cells to their present positions, and by following the patterns of speed, distance, direction and shape, it figures out the future relative positions of these matter structures. This is the main purpose of the conscious brain: to see the future through the power of its imagination by following patterns in order to figure out where your future relative position will be, or where the future relative positions of the matter structures involved will be. The difficulty in predicting the future is that you will never have all of the information all of the time in order to predict exactly what will happen in the future.

So, the future is not real; it is only imaginary. This constant recording of the present relative positions of matter structures by the memory cells, and the constant comparison of these memories to the present relative positions of matter structures by the imaginary power of the brain, tells us the speed (speed is relative motion) of matter structures as they change their positions relative to us or to other objects, and therefore gives us the illusion of time. This illusion of time is necessary for the brain to predict the future by measuring the speed at which matter structures move through space as they change their relative positions.

The more quickly matter structures change their relative positions, the faster the flow of time becomes, since you need to activate and compare more different memories at a faster rate. If matter structures stopped changing their relative positions, the illusion of time would also stop, since time is only relative to the matter structures involved.

We know that matter structures change their relative positions only by recording their previous positions in sequence in our memory, as a result of which, time flows. And this is true. When we sleep there is no recording of present events or the positions of matter structures, and we lose the concept of time. We do not know what time it is or what our relative position is. People waking up from a coma face the same situation. They don't know what the present time is or where they are, since their brain stopped recording events while they were in a coma. People with brain damage or a disease that affects their memory do not have the concept (the illusion) of time. Therefore, they do not know where they are, because time is all about the relative positions of matter structures.

A person without memory cells, or who does not have the power of imagination to analyze and compare memories, or who has a bad connection between the imaginary power of the brain and its pool of memories, won't have a clue about his identity, location, or relative position in our universe. Furthermore, he won't have any concept of time. He won't have the illusion of time going forward or backward. He will only experience the present time, without the past, and therefore, he will not have the concept of the future, because the future is always connected to the past in the brain, and the brain gets the illusion of time flowing from the future to the present and then becoming the past.

A normal and alert brain always records events (the relative position of matter structures) as it receives information from its body's sensors in real-time (present time) in the form of electrons. This ability to record events is different from brain to brain and from species to species. In humans, the ability to record events is about twenty-five times per second, and by reading these recordings (which are information molecules that are stored in the memory cells) in sequence, the brain creates recorded motion just like a movie film. When you move the film frame by frame over a light source at certain speeds, it creates the illusion of moving pictures. The brain creates the same illusion of motion on the same principle, and with the illusion of motion inside the brain comes the illusion of time. This illusion can go forward to the future and backward to the past.

The illusion of time is always connected to the illusion of motion. The faster the motion, the faster time flows. But, motion is also relative: the faster the brain records outside motion, the slower outside motion seems, and with it the illusion of time seems to flow more slowly, and conversely, the more slowly the brain records outside events, the faster outside motion seems, and with it the illusion of time seems to flow more quickly.

Certain drugs and alcohol, fatigue and old age will slow the ability of the brain to record outside events or motion. Therefore, everything will seem to move more quickly outside and the brain will have the illusion that time is moving more quickly. Certain drugs, or the excitement of a dangerous situation, will make the brain record outside events more quickly, and therefore, everything will seem to move more slowly outside and the brain will have the illusion that time is moving more slowly. When you sleep there is no recording, and the illusion of time stops completely or goes infinitely fast, which is the same thing, since motion is the relative speed of two different objects and any motion relative to zero motion will be infinitely fast.

Hopefully, you now understand that time is not real, but is only an illusion. Time is not a delivery medium; you cannot ride on it, and therefore, time travel is impossible. The future, which is the future relative position of matter structures, is not anywhere, so there is no place to go. Likewise, the past, which is the past relative position of matter structures, is only memory, and you cannot go there because it is only an illusion. The past changed form and became our present.

The five year old child you were a long time ago is not there anymore. The child changed form to become who you are in the present time. The five year old child might exist in your memory, but there is no other world in which you are living in the present time as a five year old child.

Time travel is possible only in your imagination, and it is the main purpose of your conscious mind to see the future. The brain predicts the future based on information from the past that it stores in memory cells, and it has the ability to bring to life the past and the future by way of feeling or real illusion.

Once you understand why time is really an illusion, and that it is all about the functioning, purpose and power of the brain, it takes really a small step to understand the illusion of God. The imaginary Almighty God is imaginary because He is not real and does not have a presence. (Like God said, "Only the present time is real.") Therefore, there is a direct correlation between the illusion of God and the illusion of time. Both illusions are products of the brain, which uses its power of imagination. In the same way that the brain can create (or bring to life) the past or the future through its powers of imagination, the brain can create its own imaginary God to be the ultimate leader.

The need to follow a leader is imbedded in all social animals in the form of genes that pass from one generation to the next. These genes create an urge (a need) to follow a leader, and as a group of individuals follow a leader, they have a common denominator that pulls them together and creates strength in numbers. They have a better chance to survive, and life is all about survivability to preserve ones identity. Many animal species develop a brain as a tool for a better chance of survival, and therefore will also possess the illusions of time, space and motion. They store information from the past to predict the future. So, the more memory cells the brain possess and the better analytical power the brain has to pick the right information from the memory cells, analyze, compare, and read all this information to predict the future, the bigger a brain is needed.

The human species possesses the biggest brain in the animal kingdom, and therefore, has a larger illusion of space-time. In other words, the human can go much farther in time: backward, since he/she has a large pool of information, and forward, through the power of imagination that is based on his/her vast number of memory cells. Therefore, the human beings' universe, their space illusion, becomes much larger than that of other species that possess a smaller brain. Now, it is not necessarily better, or good, to have a larger illusion of space, since there is no good or bad in nature. It is just the way things are, and everything is geared for survival. Everything is only relative.

As a social animal, the human species has always lived in groups, and each group always had a leader. The human species carries the genes that urge them to follow a leader. A leader, who was only a little different from them since he was a human, would be stronger and/or smarter, or for that matter could even have been a lunatic with a strong urge to lead, which is also genetic. The leader would reflect strength and knowledge and would be a father figure, literally, because he usually produced the most offspring in the old days. He controlled the fate and the future of the group. The group put their trust in him and followed him, many times blindly, without question, with respect and/or fear.

As humanity evolved through millions of years of evolution, the brain doubled in size a few times, reaching the size it is today about four thousand years ago, as a result of humans using it to its full capacity (which cause genetic mutations in the brain) for communication and better awareness of their surroundings. This gave early humans a better chance of survival, with the ability of brainpower to create better tools, weapons, clothing and housing. This large human brain, which has a large memory capacity, enlarged the human conception of space-time. The human universe became larger and raised more questions than answers as to how nature works and who is behind natural forces. (The subconscious triggers the conspiracy instinct.) Without the right knowledge to learn and to embed within memory, humans tried to implement the knowledge of their small universe into the larger universe by creating imaginary leaders or creators for all natural forces and structures.

By creating imaginary supernatural leaders, they satisfied their own natural curiosity of who is behind the forces of nature, which they couldn't explain through logic because they didn't have all the right information. By creating imaginary supernatural leaders or creators, they also satisfied their own strong natural urge (meaning they followed the genetic design that represents the majority) to follow a leader and to put their trust in him, which gave them peace of mind and allowed them not to waste brain energy to really investigate and explain natural forces and structures.

Humans applied all the qualities they look for in a human leader to the supernatural leader, things such as strength, knowledge, help and mercy. In a supernatural leader these qualities can be much greater because they are imaginary, and therefore, that leader should also be treated with more respect and fear. So, at the beginning, around four thousand years ago, humans created many different imaginary gods, which were different depending on the geographic positions on Earth of the groups of humans who created them. They treated those gods with respect and fear. They sacrificed humans and animals for those gods, and even presented food to them. They built statues and monuments for them, which is logical, since those statues and monuments created a presence to which the people could relate. These trends continue to this day, as people pray in houses of worship or before matter structures like books and stones. All of these things help people feel the presence of God, since it is all about feeling and belief.

As human evolution went forward, so did the evolution of their beliefs. Some humans came to the conclusion that all those different and numerous gods must also have their own leader, as they tried to implement their understanding of their small universe into the context of the larger universe that had just opened for them as their brain became bigger. They did this by creating only one leader: the imaginary Almighty God, King of all kings, Leader of all leaders, God of all gods. And as time went forward, most humans forgot about all the small gods because they weren't mentioned anymore and were therefore not embedded in the memory cells of most humans and did not come back to life in their imaginations.

Information about the gods passed through the generations in the form of stories by word of mouth. These stories became twisted as time went forward and humans interacted with different humans at different locations. This caused their stories to become mixed, which resulted in new stories and beliefs. This continued until about four thousand years ago when humans started to use written letters. Some humans came to the realization that the shape of a letter can hold information as long as they agree on this information. They can therefore mix a few letters and create a word, mix a few words and create a sentence, and mix a few sentences to create a written story, or a book. As compared with oral stories, a book is a more reliable way to hold information from the past.

Humans started to record their daily experiences, stories and beliefs in books, and those stories and beliefs about the gods became immortalized in books, as long as the books lasted and somebody could read them. So even to this day the information in these so-called "holy books" is the source of past information about the gods that comes to the present time. But the more information the book holds, the more complex it becomes, and as people read it the information the book provides interacts with their own feelings and knowledge, resulting in different knowledge, understanding, explanation and meaning of the information the book provides, which will be different from person to person depending on their experiences and genetic makeup. In other words, different people will interpret the information in the book differently. Furthermore, the same person will change the way he thinks, the way he views and judges things, many times throughout his life, since his brain constantly gathers new information and erases some old information as time goes forward.

So, all in all, the imaginary God manifests Himself based on the information that humans acquire throughout their lifetimes and especially throughout their childhoods, and is fueled by the natural urge and feeling of humans to follow a leader.

A child born in Ireland will create a different imaginary god than a child born in Pakistan. Each imaginary god is based on information that is stored in memory cells, and both children will have received different information based on their surroundings. Each child will create a different god in his or her imagination, concerning how this god looks, what he wants from them, and how they are meant to follow and worship him. Both children will believe that their god is the right one, since he is the only one they know. So, the imaginary God is the creation of the human brain, and is based on the information that the brain stores in its memory cells.

And here is the correlation between the illusion of God to the illusion of time. They are both based on the ability of the brain to hold information from the past and bring it back to life with the imaginary power of the brain. Without these two things (the information molecules that are stored in memory cells and the ability of the imaginary power of the brain to read this information in the right order) there is neither the illusion of time nor the illusion of God. Both illusions will seem very real since it is the main function of the brain to bring memories back to life, back to the present time, with its imaginary power. Therefore, the illusion is very powerful and seems real, with two differences between the illusion of time and the illusion of God.

First, the illusion of time: The normal and functioning brains of all species that contain memory cells possess the illusion of time, which is the ability of the brain to recognize the motion of external matter objects and to predict the future relative positions of these matter objects by reading the past and comparing it to the present. And therefore, the brain gets the illusion that time flows from the future to the present and then become the past. The brain can see, feel or predict the future by way of this illusion, and orders the body to move or respond accordingly, either to engage or avoid the future as it sees fit. The bigger and more complex brains that hold a vast amount of memory (like most brains belonging to the mammal family, including humans) will also create more choices of possible future scenarios and can look farther into the future. The human brain, which is the largest in the animal kingdom, possesses the most memory molecules in its memory cells, especially in the long-term memory cells, and can therefore go much farther into the future (the future as the brain sees it). But do not forget the future is not here. It is not real. It might be the way you predict it, and it might not. This depends on whether you have all the right information and have also analyzed it correctly, all according to the laws and possibilities of our universe.

But the human's brain can also hold abstract information, imaginary information, or in other words, information that is not based on reality. The interaction of this information with natural feelings and urges that come from the subconscious can result in predicting a future that is unrealistic. The human's brain can order the body to respond and move toward this unrealistic future the same way that it predicts and responds to any other future. For example, some people will predict that there is a heaven in their future, a belief that is based on information they receive throughout their lifetimes from other people in whom they put their trust. This information could also come from so-called "holy books" that tell people how to act and behave in order to reach that future, and the people act and behave according to this information to get to the desired future. It could be by praying to an imaginary god, being polite and helping other people, or even through extreme acts like killing other people or killing themselves. For those people, imaginary information that will result in an imaginary future is real, just like information that is based on reality, since the imagination is very powerful. Who you are is based on your imagination, which is based on the information you collected throughout your lifetime. And the interaction of this information with your own natural urges and feelings, which come from your subconscious, creates your identity and your universe, which is unique and different from person to person and from time to time.

Secondly, the imaginary God (imaginary leader) can be created in the brain of all species that have a brain which contains memory cells and have the genes that urge them to follow a leader. All these species, including humans, will create imaginary leaders in their brains who are based on reality. In other words, all these species hold information in their brains from interactions with the real leaders of their past, and create memory molecules in their brains of what the leader looks like, his smell, his voice, and also what this leader wants from them. They store these information molecules in their memory cells and can bring these memories back to life with their power of imagination. So, the leader is imaginary. This leader is created in the brain of all these species, but is based on real information. Once they encounter their real leader either by sight, smell or voice, as the information comes in real-time through their sensors, they recognize a match between the imaginary leader and the real and present leader. They act and behave toward this leader according to the instincts and the knowledge of that species, which is genetic and comes from the subconscious. This also comes from the knowledge that the conscious mind possesses in its memory cells in the form of memories of what this leader wants them to do.

For example, a dog is a social animal because he has the genes that urge him to follow and please the leader and also possesses a large brain that contains many memory cells, so he will remember his leader by storing these memories in his memory cells and can bring these memories to life through the power of imagination. And therefore, he has an imaginary leader. The imaginary leader in this dog brain could come to life in his imagination if the imagination were triggered by a dream or outside event of sight, smell or sound. Once this dog encounters his leader and recognizes him as the leader by sight, smell or sound, as there is a match between his imaginary leader to reality, he will behave and act toward this leader with the knowledge he possesses from his subconscious (collective memory) and from his conscious memory. The dog will approach the leader of the pack carefully, with his tail between his legs, and try to lick the leader of the pack. The dog leader could also be a human, whom the dog will usually approach by wiggling his tail, jumping, licking, and following him, depending on what his human leader taught him.

Human beings will have many real leaders throughout their lives, real leaders they can see and hear in real-time. But the large human brain has the ability to go much farther in time, forward to the future and backward to the past, and has the ability to follow a long and complex pattern. To follow a pattern is the ability of the brain to compare different memories that came in at different times, follow this pattern, and figure out the continuation of that pattern into the future, which therefore, gives the brain the illusion of time. So, all brains follow patterns and they all have the illusion of time. The large human brain has more memories and can follow a pattern in more detail and longer distance.

For humans, this has raised more questions than answers about who is behind the natural forces and structures that their large brains discovered as their universe size increased with the size of their brains. Without real knowledge of the natural forces of our universe, humans adopted and implemented the knowledge they already had as they recognized it in the pattern and the structure of their daily lives, and figured out that there must be a creator or a leader that is controlling the forces of nature, and so, the imaginary leader, or God, was born. God is a purely imaginary leader that nobody can see or hear in real-time, and therefore different people will see him differently in their imaginations, depending on the memories they receive up to that point in their life. (My God of Knowledge looks like a luminous old man who speaks English with a heavy NY accent and smells like the forest. What does your God look like? What do you mean by saying you do not know what your God looks like? How would you recognize him if you saw him?)

So, following patterns and comparing the relative positions of matter structures gives us the illusion of time and also the illusion of God. This illusion of God gave us an easy way to explain the forces of nature and the creation of our universe. It is much easier to blame somebody else for something you do not understand, and it also gives you a reason to not waste brain energy to follow a pattern. Science is all about following patterns and understanding the logic behind them. If you understand the laws of nature and apply these laws to the motion of matter structures, you can see the patterns and can go much farther into the past, all the way to the creation of the matter in the Big Bang about fifteen billion years ago (not five thousand years ago when God was supposed to have created the world). We are able to follow the pattern of the chain reaction set in motion by the Big Bang with the ability to read and understand in detail matter structures as small as atoms, subatomic particles, and even light waves, which are all memories, or in other words, information from the past. Because, when we understand the past, we are able to predict and design the future much better.

The imaginary God is purely the creation of the large human brain that wishes to have Him in its life, as humans feel an urge that comes from their subconscious to follow a leader. The human imagination can create a supernatural leader with supernatural powers to help them to bypass reality and to give them peace of mind. The illusion of time helps all species that have a brain to see and create a better future, and it serves this very purpose in the human brain. But the illusion of God in the human mind can disconnect humans from reality, and can create more problems than it is worth. Depending on and believing in a supernatural leader can blind us from seeing the real future that is being created by our present behavior.

As I said at the beginning of this chapter, once we understand the illusion of time, we will understand the secret of this universe and we will understand God much better. And this is true. Now we know that the universe is recreated inside of the brain, and that the way the brain sees the universe depends on the knowledge the brain has and stores in its memory cells. The human brain creates God by reading memories from its memory cells, and combines them with imaginary power under the influence of the subconscious. The human brain creates the illusion of time by recreating the outside motion of matter structures inside the brain. The byproduct of the recreated motion in the brain is the illusion of time, and with it the human can choose his future, since the future is open to a degree, and this depends on the knowledge and the energy the human can harness.

Animals with normal brains also have the illusion of time and can imagine the very near future, but they do not have the concept of our imaginary God. Even though they follow their own leaders (or anything that resembles a leader) in real or present time, animals still do not have our imaginary God because their brains do not have the ability or the capacity to hold abstract information.

So, the God of Knowledge was right when He revealed Himself to me in the holy mountain of West Virginia. Animals that have a brain still do not have the concept of our God. Only humans, with their large brains, can create the imaginary God, as long as they wish Him into their lives through their brain's power of imagination.

This makes the imaginary power of the brain the most powerful force in our universe. It is our fifth dimension, and with it we can go far into the future, far into the past, and to faraway places. We can create imaginary structures, we can create the illusion of time and the illusion of God, and with this power we can leave our universe, a material universe, a matter universe that created our brain and the imagination.

33. A Supplement for God and the Illusion of Time

This chapter is for all of you who do not understand or comprehend that time is not real, but is only an illusion. You are not alone. Even Einstein, with his great mind, didn't realize that time is not real. Time is not the fourth dimension. The motion of matter structures is the fourth dimension.

But if time is not real and is only an illusion, how can an illusion have laws? For example, the first law of time says that everything has to have motion for time to flow. The second law of time says that if there is a beginning of time, there must be an end of time. The third law says that the end of time will resemble the beginning of time. The fourth law of time says that matter structures change forms; it is also a hallmark of time.

But as God explains it to me as I write this chapter, our universe does not have such a thing as "**TIME**". What it does have is motion. The brain can see and feel this motion by recreating the outside motion of the surrounding world inside of itself. But, to put an order to the relative motions of the many different matter structures of which our universe is made, and to put an order in the sequence of interactions of matter structures as they change forms or relative positions, the brain creates the illusion of time. This happens in any normal and functioning brain, in all animal species that have memory cells to record the outside relative positions of different matter structures.

The brain can record the outside motion of matter structures from the present time or current position only, and as those matter structures move their relative positions change and the brain can compare the old relative positions from its memory cells to the current relative positions. It knows there was a motion, it knows where this matter structure came from in the past (it is not there anymore: the illusion of time), and it knows where this matter structure is heading in the future (it is not there yet: the illusion of time). All brains create the illusion of time, since the main purpose of the brain is to gather information, remember it, and use it to recognize external matter structures by comparing them to the information it has in its memory cells. And if it is aware of movement, it can predict the future by creating the illusion of time. The brain performs all these functions naturally from birth, without thinking, with minor adjustments at the beginning of life, as the brain gathers information and stores it in its memory cells and learns by itself through trial and error, since it is designed for that purpose by the genes.

So, the brain creates the illusion of time naturally, without thinking, just by comparing the relative positions of matter structures from its memory cells to the positions of those same real matter structures as they are outside in the present time. If the brain can see that the position is different, it knows there was a motion, and with the knowledge of that motion comes the illusion of time.

We always use the motion of matter structures, as they change their relative positions, to get the perspective, or the illusion, of time. A very simple example is the clock. We know that one hour has passed when we see the hour hand pointing to the number five, and we remember that it was pointing to the number four before, which means it has changed its relative position. As you know, twenty-four hours is one complete spin of Mother Earth around her axis relative to the Sun. So, relativity is at the root of the illusion of time, and all brains compare relative position naturally. It took a giant of a man like Albert Einstein to point out to us that relativity is at the root of our universe. Everything is relative, nothing is absolute, and this is true also for the illusion of time, which is based on the relative motions of different matter structures. Everything is relative to your brain as the brain recreates the outside world inside of itself.

Time is also relative to your own time. For example, one year for a five year old child is twenty percent of his lifetime, and therefore will seem to him as a long time period. But for a fifty year old man, one year is only two percent of his lifetime, and therefore will seem to him as a much shorter time period relative to the five year old child. (Try not to pass the hundred year mark, when one year would be less than one percent of your lifetime and time would zoom by you very quickly.) Time will also move very fast if you keep yourself busy, or if you do not pay attention to outside motion.

If you can create the illusion of time, it is possible for you to create the illusion of God, because both illusions are based on memory cells. So, is God real? Is yesterday real? Yesterday does not have a present, so it is not here anymore, and it is not real. It is only memory. And certainly, my yesterday is not like your yesterday, since we have different memories, and without memories, there is no yesterday; there is nothing.

Now do you remember what God said in the last chapter? He said: "Only the present time is real, and it is motionless." If you read between the lines, you will know what God meant. There is no "real" time, since our universe is in motion all the time. Therefore, there is no such thing as "**time**" or "**present time**" to hold on to. As you know by now, knowledge itself involves the motion and interaction of matter structures, and therefore, without motion there is no knowledge.

God answered the questions I raised earlier in this chapter about the laws of time by saying: "The awareness and the knowledge of these laws by the brain helps the human brain to cope with reality as the outside conditions are constantly changing due to the motion of matter structures."

The knowledge of these laws creates the illusion of time as a tool for surviving longer, by enabling the brain to predict the future. This is just like religious laws, which supposedly came from the illusionary God and are mostly a good common sense approach. They are designed to prolong life, bond people together, and put an order in the daily life of the tribe.

First, there is the law of time which says: everything has to have motion for the illusion of time to flow. We know that the illusion of time is triggered by our knowledge that matter structures are in motion. So what happens when we are in a closed, silent room in which nothing we can see moves, including us. Does the illusion of time stop? The answer is: no. We still are aware of the movement of our own body, which is made from many structures. We can feel the motion or the rhythm of some of these structures, such as the heart and lungs. Furthermore, our consciousness, our awareness, is open. (Awareness is the movement of electrons, atoms and molecules inside the brain, and especially the movement of memory molecules and the interactions among them that create knowledge, thoughts and feelings.) Therefore, we still get the illusion of time flowing forward as the brain puts an order in the movements and the rhythms of its own body by sequencing their numbers (like what comes before what by memorizing them in that sequence).

The second law of time: if there is a beginning of time, there must also be an end of time. We know that time is always connected to matter structures. Time starts when a new structure is created, and time stops for this structure when it is not there anymore, since everything in our universe has a lifespan. For example, a baby is born and time starts for him. He wasn't there before. He didn't exist before. And as this baby becomes an old man and dies, time stops for him. So, time does have a beginning and an end, just like the matter structure to which it is connected as a reflection.

The third law of time: the end of time will resemble the beginning of time, but will never be identical. If you know the second law of time, it is easy to understand the third law of time. You know that time starts with the creation of a new matter structure, but that this new matter structure has to be put together by some energy that moves the matter structure to create a new and different matter structure. As this matter structure disintegrates, time stops for this matter structure because it is not there anymore. So, to put a matter structure together is just like disintegrating it, only backwards, just like in a movie film when we play it in reverse. So, the beginning of time starts with the creation of a new body structure and stops when this body is disintegrated, and the disintegration (end of time) will resemble the creation (beginning of time). This law of time is very important since it lets us see the future (the end) as long as we know the past (the beginning). God and I will give you a few examples. Mother Earth was put together by many collisions of heavenly bodies, which started about five billion years ago. The end of Earth will resemble the beginning: it will end also in a collision with some of those heavenly bodies.

Our universe started from a different dimension and time started when our universe created matter. Time will stop as all the matter in our universe eventually becomes a different dimension.

A baby is put together in his mother's womb, and when he is born time starts for him. He is born alone with nothing. (Even if he is born as part of a set of conjoined twins, he is still alone as an individual, and we are all individuals, each a force of one.) And eventually, when he dies as an old man, time stops for him and he dies alone with nothing. His body will disintegrate, ashes to ashes and dust to dust. He came from Earth and he is going back to Earth. Even if he dies in Space, there is no violation of this law of time, since we really came from Space and eventually we all will return to Space. So, this old man would just be taking a shortcut.

Nazi Germany was born as a cult of terror and violence, with destructive political and belief structures, and it eventually died the same way. The end will resemble the beginning, and just like God said at the beginning of this book: "If you start terror, you will live and die in terror and violence."

The fourth law of time: matter structures changing form is also a hallmark of time. When we look at a matter structure and it does not look the same as before, we perceive that this matter structure got older since we remember what it looked like before. The illusion of time is triggered by comparing memories to the matter structure's relative position in the present time. For example, when a baby changes form to become a child we relate memories of this child to when he was a baby, and we see that he changed form to become a child. This gives us the illusion of time: the baby became older. This baby will constantly change form to become a boy, a teenager, a man, a middle-aged man, and then an old man, and all the while we will perceive him as getting older. But the illusion of time can go backward also, according to the fourth law of time. If we see a matter structure as it used to be before, the illusion of time will go backward. For example, if this old man reshaped his body through exercise, the right nutrients, genetic therapy and/or plastic surgery, we would perceive this old man as younger than he is, as though he went backward in time, because we remember him as older.

So, all in all, there is no time. There is motion and there are memories. The brain has the ability to compare relative positions of matter structures from its memory to the relative positions of the same matter structures in the current (present) position as it detects movement or detects change. This is what we call **TIME**. If there are no memories or movement of matter structures, there is no illusion of time.

In theory (it is really impossible), if we take a human body and preserve it in such a way that we can stop all of its movements, all the way down to its molecular structure, this human body will stay the same for eternity. It won't change forms and it won't get older or younger.

If we travel in a spaceship at the speed of light we could be anywhere in our universe in no time, since our body stops all of its movements, all the way down to its building blocks of matter, and time stops for us. However, for people on Earth this spaceship still needs time to reach different points in our universe, since it traveled only at the speed of light.

We are all of the same age. We are all made from the building blocks of matter that were created in the Big Bang about fifteen billion years ago. We are just put together at different times.

34. Evolution

Evolution is simply the name of the process of changing the size and the shape of matter structures through the forces of gravity and energy over time. It is the recording of events imprinted in matter structures in the form of size and shape.

This also applies to the change in the structure of any life form and any manmade product and design, including religion, science and belief structure.

As you see, evolution happens everywhere and all the time. However, it is hard for the human mind to believe in evolution since it happened in the past, whereas in our short lifespan we do not see or notice these changes and we really do not want to spend the brain's energy looking for them. It is the way of nature to conserve energy, and it is much easier to blame somebody else for the creation of life, such as **UFOs, intelligent design** or **God.** But if you use your brain to pursue knowledge of the way nature works, you can see the pattern, you can see evolution.

For example, a body of water streaming downward with the help of gravity seeks a low point in the ground, and if there is a steady supply of water it will create a stream of water, a river stretching many miles with a dead end at its lowest point, which is usually a big body of water, like an ocean or a lake. Now, this river depends on its water volume and its speed to give it energy to curve a path in the ground, seeking a straight line and the shortest route to its dead end. But, the straight path is full of obstacles, structures in the form of hills, mountains and rocks, whose energies are greater than the energy of the water and therefore will not move. So, the river has to go around these obstacles, swerving left or right and sometimes both ways to bypass them. And as time goes forward, some of those structures, or parts of them, will give up to the constant friction generated by the energy of the water and will move or break down. Some paths of the river will get clogged with sediment and debris resulting in a change of the path of the flowing river.

This constant changing of the direction and the size of the river is the evolution of the river, because the river's shape always depends on other energies and structures, be they manmade structures, earthquakes or water supply. There are always changing conditions because everything is alive, everything is in motion, and everything is going through evolution and changing through time.

The stones in the riverbed are going through a different and accelerated evolution compared to their counterparts on land, simply because the river's water carries more energy to change their shape. The constant movement of the water moves other stones and debris to give the stones their round shape and smaller size, and these stones, with the help of the water movement's energy, hit other stones and rocks to create ever smaller stones, and so on, all the way down to sand. So, by looking at the stones and the sand in the riverbed we can figure out the evolutionary stage of the river, where it was flowing before or how big it was, and how fast it was flowing, and we can determine how old it is, since evolution and time go hand in hand.

Evolution is the change in any structure over a period of time, and time is the changing relative position of matter structures. In other words, evolution, or time, needs movement and motion of structures. If there is no motion, there is neither evolution nor time. Time (motion) is a fundamental law of nature, and so is evolution. Both time and evolution are byproducts of the movement of matter structures. Therefore, for humans to perceive this, we need to activate memory cells to compare memories and to detect changes, or movements.

The relative positions of matter structures change all the time, and so does evolution. The structures within each structure change their relative positions all the time because everything is in motion, fueled by the play of gravity and energy.

Therefore, evolution in principle is always a byproduct of the chain reaction of objects with their surroundings, as matter structures including life forms here on Earth evolve to become different from their predecessors. That means that evolution is a fundamental part of nature. It is always there.

Time and evolution can be rewound. They can go backward in different small structures with the help of energy to move these different structures back to the same position they were in, let's say, ten years ago. So, time will go back for these structures ten years, and so will evolution. In other words, it is now evolution in reverse, but nevertheless, it is still evolution. For example, if we build a new house and live in it for ten years, the house will have its own evolution. The paint will lose its original color and in some places it will come off. Some of the walls will have some cracks and some of the house structures will fall off or break. The floor will have signs of wear and tear, and after ten years the house will look like it is about ten years old.

But what if you can fix the house and put everything back to its original condition by using energy? Time will go back for the house since all of its many different structures will have the same relative positions to each other as they did ten years ago. Time will go back ten years and the house will be new again, and time will tick again for the house as long as the house's many different structures change their relative positions. The same thing will happen with the evolution of the house, which will go back to when it started ten years before and restart again as long as there is movement of the many structures of which the house is made. So, evolution, just like time, can and does go forward and backward when a structure goes back to a previous position, which it does always with the help of energy. And likewise, it can stop if there is no movement of structures.

More about evolution in the next chapter.

35. Conclusion

Our universe contains a huge number of structures that can also be referred to as identities or information units. By way of their motion and the way they interact with their surroundings, we can read the information of their evolution: where they came from, where they are going, what they mean and how old they are. But to be able to read this information you need to follow a pattern, you need knowledge, just like you need knowledge to read this book. If you do not understand the structure of each English letter (the information unit), then you cannot get the information the letters represent, and therefore you cannot get the information from this book, but it does not mean the information is not there.

Evolution happens every day, every second. We can notice it over a long period of time because the changes in a structure are more profound, but if we look hard enough we can see evolution everywhere and all the time.

Bacteria can change their structure in a matter of months to adapt and survive lethal medicines like antibiotics. Because they multiply so fast (in minutes or hours), they can produce many generations in just a few months. Some of them adapt to medicines by changing some of their structures, and as some of the bacteria genes mutate, as they interact with their new surroundings, they evolve into drug resistant strains. The same occurs with viruses, which also reproduce rapidly, with thousands of new generations in just a few months. Some of the new generations of viruses can go through evolution and change their outer structures to adapt in order to attack different cells of a living body or cells of a different species altogether.

Our body goes through evolution every day: it gets older. It changes structure all the time. In principle, getting older and evolution are the same thing. Many humans look a little different from other humans of only a few generations ago. They have on average become taller and wider because conditions have changed with time through the evolution of agriculture and transportation. This means that there is more food available with more variety year-round, more than ever before. People in general live longer and have more comfortable lives than their predecessors, thanks to advances in medicine and science that have gone through accelerated evolution in the last few generations. Science can accelerate evolution much faster than nature (which works by trial and error) by putting different combinations of structures together with the help of energy and the knowledge of nature's laws.

The gene structures in all life forms change very little in one generation and need many generations for change to become apparent. That means it can take many millions of years for profound changes to occur to increase diversity in the larger and more complex life forms that multiply much more slowly than bacteria and viruses.

Humans evolved from the ape family. There is a very close resemblance between us and apes in the way we look, in the shape of our bodies and in our behavior, and we do share the most genes with them. We resemble apes, but we are not identical to them. Just like we resemble our parents but are not identical to either one of them, and that is only a difference of one generation. Now, imagine two thousand generations, or thirty-two thousand years backward in time. (A generation is the timespan from when you are born until you create your first offspring, which in the old days was about sixteen years on average.) In this many generations, the difference between how we look now and how we looked then is much more profound. We did look like the cavemen. Now, imagine two hundred thousand generations, or three million two hundred thousand years ago, when we looked like the apes from whom there is no doubt that we came. We share the same ancestors, the same parents, if you will. All that was needed was a single ape (they are also God's creatures) with mutated genes to pass his or her genes to the next generation. And this is precisely what started the course of human evolution.

Nature has left us many clues about the evolution of life and its diversity on Earth in the form of the structures of many life forms and/or their chemical prints. But it is against the laws of nature to preserve every structure of life form that was ever developed through life's evolution on Earth, from billions of years ago, or even a few days ago, because everything changes forms, sometimes beyond recognition. This is what happens when creatures get consumed by other life forms, or with the effects of weather and chemical interactions. Humans evolved from other species and owe their existence to the many different species that went through life's evolution in the span of billions of years to adapt and survive different conditions and the competition for space and energy.

All the life forms on Earth are related, just like all the matter in our universe is related. All the matter in our universe is made from the building blocks of matter, the pairs of gravitons that were created in the Big Bang about fifteen billion years ago. All life forms on Earth evolved from the first gene that was created on Earth about four billion years ago. All life forms on Earth are made from many genes. These genes are mutated genes of the first gene, which mutated to build a body that is suited for their previous or present condition.

Throughout Earth's history many species evolved and disappeared to be replaced by other species. Some species survived only in small numbers and went through evolution to adapt to new conditions. After many generations they changed so much that they did not resemble the species from which they came, and only their DNA can tell us the connections. This process of species splitting into groups that face different conditions, and therefore change the course of their evolution, is the reason behind the explosion of so many different life forms. There are millions of them on Earth, and they are all related. The more genes they share, the closer the relationship is, and the less genes they share, the farther apart they are in time and in the directions of their evolution.

All life forms compete for energy and space: species against species, groups against groups within the same species, and individuals against other individuals within the same group. But life is not all about competition. Life is also about living in harmony and sharing. You give me something, and I will give something in return. Business was invented by nature long before humans discovered it, along with recycling and efficient use of energy and space, because everything is limited.

The plants use carbon dioxide from the air, water and nutrients from the ground, and with the help of the Sun's energy, they produce oxygen, carbohydrates and other nutrients for the grazing animals. In return, the grazing animals consume oxygen and carbohydrates and produce carbon dioxide that they put back in the air for the plants to reuse. And animals help disperse the plant's seeds to faraway places and help to pollinate them. Predators live off of the grazing animals, and in return for the food supply they keep the grazing animals in control, not to outpace their food source. In doing so, the predators protect the plants from overgrazing. The predators also eliminate the old, sick and unfit grazing animals, and by doing so help the grazing animals to maintain a good gene pool and remain disease free.

Our body, like that of other animals, hosts many different families of bacteria. In return for shelter and food, these friendly bacteria help to protect the body from other harmful bacteria and help to digest food that would otherwise be wasted energy.

This ecosystem on Earth has been in balance for billions of years. It will remain in balance as long as there is energy coming from the Sun and Earth preserves the right conditions (more or less) for its life forms.

This balance among the life forms on Earth remained steady as long as no one species had a big advantage over other species in their ecosystem, until humans arrived as a split group from the apes a few millions years ago with the advantage of the big brain. As time went forward, their brain got even bigger in size as a result of using it to its full capacity in order to communicate and to understand nature. Humans used their brainpower to build extensions to their bodies in the form of clothing, weapons and tools to better their chances for survival, and they spread all over the world with all its diverse conditions.

The large human brain gave rise to explosions in all the departments of science in the last one hundred years. It gave rise to the ability of humans to build extensions to their bodies, to develop wings and cross the globe in a matter of hours and even leave Earth to enter outer space, touch the Moon, and come back. Extensions to their legs in the form of motorized transportation enable humans to haul almost anything to great distances in a matter of hours or days, instead of walking on their own legs for many days or months. The humans built computers as an extension of the brain, to hold and communicate information or calculate future events. Medical science breakthroughs have enabled humans to replace body parts (which no prayer can do). Humans developed medicine as an extension of the immune system, to ward off disease, and by doing so have extended life expectancy to an average of seventy-five years instead of forty years a few generations ago. There is also much less infant mortality than there was a few generations ago, and that means more people reach adulthood, produce more offspring and live longer than ever before. And that will create many problems.

The human population on Earth was only a few million fifty-thousand years ago, now it is over seven billion. Without containing the human birthrate there will be a population explosion. The urge to reproduce that comes from the subconscious is very strong, and it is also fueled by many religious beliefs, beliefs that were created a long time ago at a time when the conditions were different. Having many offspring was a good idea because many did not survive to adulthood. The larger the family or the group, the better the chance for survival and the preservation of the identity of the family or the group. But today the conditions are different. People live much longer and quality is more important than quantity. Explosion in the human population means we need more space, space that belonged to other species and plants, and this means less and less habitat for them, which means changing the balance of the ecosystem with costly consequences.

Earth's resources, which are limited, will get depleted much more quickly. We will need more energy and in the future the overall quality and the value of life will go down as a result of an explosion in the number of humans. Overpopulation will create friction and many problems between the people who have energy, knowledge and resources, and the people who do not have enough. Life will be more complex and more difficult. Many people will turn back to religion, which was created a long time ago under different conditions and under the massive influence of the subconscious.

This old religions believe in an absolute, and there is no absolute in nature. Everything is only relative. Religions are shortsighted in predicting the future and are intolerant and disrespectful of other individual humans and other religions. They promise better ways of life and a better life after death.

Yet, we are dealing with life here on Earth, and only the awareness of the conscious mind, with its ability to see far into the future and its understanding of nature's laws, can control the subconscious. With the awareness and knowledge of the conscious mind we can control birthrate to have better quality of life and longer life. With the knowledge that comes from the conscious mind we can harness the Sun's clean energy directly in order to produce electricity and to split water into hydrogen and oxygen for our use. We can occupy the Moon and some of the planets in our Solar System, along with the empty spaces in between and perhaps even other solar systems. In doing so, we can expand and bring the evolution of life that started on Earth to outside worlds. We can do that for which we are designed: expand, preserve and spread our identity. We can accelerate life's evolution and humankind's evolution through science, by building better extensions and replacement parts for the human body with better designs than nature. In just a few thousand years, some groups of humans could occupy faraway places in deep Space, independent from Earth, and those different groups of humans would have a different direction of evolution, depending on their knowledge, their conditions and gravity force, and may look and behave differently from other groups of humans.

Just like everything else, all humans are not created equally. They look different from each other. They each have a different brain structure and different information in it. So, everybody is unique and a little different from other humans, and therefore, we are diverse. There is no one human that is superior to another. Everybody has a function, and together all humans make a group that functions like a body, with the many different and diverse parts that make the body the way it is, or in other words, that make its identity.

Life on Earth started by taking diverse atoms and molecules giving them energy, the right conditions and freedom of time over millions of years, allowing unique structures to pop up (life) because it is possible.

So, freedom and the right conditions do help evolution to accelerate, just like democracy gives freedom and incentive to its people to excel in many fields. Free trade, free speech, freedom of thinking, diversity, and free competition also help to accelerate evolution. However, total freedom is not for everybody. There are some people who are controlled by their subconscious and will try to take control of other people and actually try to go backward in time, backward in evolution, and will unleash the dark side of their subconscious with its shortsightedness.

The subconscious is an important and integrated part of us. It gives us the will to survive, all our instincts and all our feelings. Without the subconscious, we would lose our emotions and life would be boring without the feelings of joy, love and sadness. A human would be just like a worm, a plant, or a computer, that lives its life just the way it is because it does not know any better. But the conscious mind, with its ability to calculate information and see far into the future can help the shortsighted subconscious to plan ahead and survive longer by giving it information and awareness, and by controlling the urges that come from the subconscious.

The subconscious and the conscious mind are made from a huge amount of small matter structures. Matter structures are also information units, so our knowledge, behavior, and the ability to think are limited and can represent only these matter structures, the connections and the interactions between them. Although a huge number of possibilities and combinations exists for matter structures, they are still limited by numbers and time, just like our universe, which is limited by its four dimensions, its size and its lifespan.

So, God was right again: we are all robots. We are made from a huge number of similar matter structures (identical structures will create the same information and behave the same way under the same conditions), and therefore, we are all about the same. We are each programmed by our structure, which was built by the genes to do what we do, nothing more and nothing less. Our structure changes with time, through the evolutionary process, and with it changes our way of thinking, which is a byproduct of our brain structure.

We are each a product of the evolution of our universe, which transpired through a complex series of matter-energy chain reactions of a huge number of small structures that formed bigger structures of life forms on Earth. We are one of many life forms on Earth, shaped and designed by the evolution of life to survive as individuals and as part of a small group that is part of ever larger groups. The human is like a sophisticated robot that redesigned itself through evolution to survive, by looking for an energy source, replicating itself, and rewarding itself with the feeling of pleasure when it succeeded in doing that for which it was designed.

There is no absolute and real God, just like there is no real and absolute truth, since everything is only relative to your brain. We are all alone. We are controlled by nature, which is our universe with its four dimensions. We are a small part of nature. We can control nature in a limited way for a while, but nature is much bigger and stronger than we are, so it will win in the long run to follow its orders and change forms. Nature treats humans just as it treats any other structure in nature. There is no special treatment for life forms in nature, and we are just another structure that depends on its energy to survive and to preserve its identity. Everything has a reason and basically everything is a result of the play between the forces of matter-energy and gravity. These forces created our past and present and will create our future.

The good news is that there is a God and that he is in our imagination (There is no God but the imaginary God). If we can imagine God, that means we have an imagination. Imagination is our fifth dimension, and with it we can disconnect from our world, our universe, and we can be at any place at any time. As long as we have imagination, we can disconnect from our bodies, disconnect from matter, disconnect from time and create different worlds, different dimensions and different gods to lead us and show us the way to a better life, to an immortal life.

But in order to imagine, we need a structure, a real structure made from matter and fueled by energy, just like our conscious mind with its memory cells that hold a huge amount of information. And to trigger the imagination we need outside events, information molecules that we receive through the body's sensors, along with the urges and feelings that come from the subconscious, or from collective memory. These are also molecules, and the combination and the interaction of all or some of these molecules in the conscious mind will trigger the conscious mind to imagine. The subconscious will direct and reward the conscious mind with the feelings of happiness, joy, fear or anxiety, among others, for its imagination, just as if it were real.

So, the imagination is really limited to the sum of the information molecules involved, and it is different from person to person because we all do not have the same exact brain structure and the same exact information, and also do not have the same exact subconscious (collective memory) because we all came from different places. Two of us never occupy the same location at the same time, and therefore, we do not receive the same information. Information also changes from time to time, and so do our imagination and beliefs.

Throughout human history humans have perceived God differently depending on their location and time, and we can see these differences through the evolution of religions and beliefs. At the dawn of religious belief, there were many different gods, and later on most groups of humans consolidated all the gods into one Almighty God. Then again, different groups see their God in their imaginations differently, in terms of what He looks like, what He wants from them, and how they should please and worship Him. Again the perspective of God is different from person to person within the same group, and from time to time within the same person as the person receives new information and erases the old information. So, our imagination, our thinking, is the reflection, or the reading, of these connections, and the interaction of these certain information molecules in our brain in the present time.

These information molecules are made out of atoms, which themselves are matter in a certain order that represents information identity that is subject to the laws of our universe with its four dimensions. That means that nothing is forever. Everything needs energy, everything has to flow, everything is a result of a chain reaction, and everything has to change forms sooner or later to justify the law of time which says that all matter has to be in motion and must go in a full circle. If it had a beginning, it must have an end, and the end will resemble the beginning. Matter was born from a different dimension and will end in a different dimension. When all the information our universe contains in the form of the relative positions of matter structures vanishes, it will disappear to become a different dimension and our time will stop.

We are prisoners of our universe, and if you cannot beat it you should join it, by understanding nature's forces, not by following the path of superstition and worshipping mean and blood-thirsty gods. You can create and follow a loving god (what the world needs is love).

Happiness comes when you fulfill your body's design, when you fulfill the urges that come from your subconscious. This includes the urge of curiosity (the urge for knowledge), which stimulates the brain by increasing awareness and understanding of what's going on around it in nature. The brain is designed by the evolution of life forms on Earth to give an advantage in the struggle for survival to the life form that possesses this urge to understand nature. Stimulation of the brain by this urge helps it to grow, to excel in many different directions, and to accelerate the evolution of life.

The evolution of life forms on Earth is one part of the evolution of our entire universe, with its many diverse ways of playing the game of life.

Our universe is a matter universe, and that means that everything is connected to matter. Our bodies, our feelings, knowledge (God) and time itself are all connected to the relative positions of matter structures and the way they interact among themselves.

True or false? Let me know.

36. Reincarnation

As a religious atheist, I believe in reincarnation. Everyone has their own beliefs and visions of what will happen to them after they die. The beliefs or the visions of reincarnation are created in the human brain and are based on the information a person receives up to that point in their life. These beliefs also come from the deep feeling in their subconscious that they will live forever, or that at least their soul or their way of thinking will survive. This information is embedded in the human brain in the form of matter structures in a particular order. Our beliefs and our way of thinking are the reflection, or the reading, of these molecules inside the brain. This means that beliefs, vision or thinking can be changed, twisted or destroyed as those matter structures inside the brain change their structural order due to physical changes that occur over time. This may also happen by way of collision with other matter structures, chemical reactions, radiation or flow of electrons.

These beliefs or visions of reincarnation can stay or change inside the human brain regardless of what really happens after death. These beliefs are different from person to person because we are all a little different from each other. These beliefs also change through time within the same person as the conditions and the information changes.

But let's leave this serious talk behind, and let's have some fun.

When I die (and I will make sure that I will be there to supervise), I would like to give my dead body to an alligator farm to be fed to the alligators. I always wanted to know what it would feel like to be swallowed by an alligator and come out the other end. I hope I won't feel any pain, and I do not think that I will be scared. I will be eaten by a few alligators who will fight for my remains, and I hope they will enjoy their meal. Some parts of me will stay with each alligator, and I will become an alligator, and some parts will come out the other end and will be eaten by bacteria, and so I will also become many, many bacteria, and will have a boring life of eating dirt and reproducing by dividing without having sex.

The part of me that becomes an alligator will have more fun. I will finally be able to eat rotten meat without getting sick, swim in dirty water and fight other alligators for food, territory and to make love to other alligators. And when the time comes, I will be slaughtered, and my skin as an alligator will make brand name belts, shoes, and other leather products, and I will travel all over the world with the people who bought these products, sometimes in first class without buying an airline ticket! My alligator meat will be served in gourmet restaurants and will be eaten by many people, and I will become many different people from different backgrounds, male and female alike, of all ages, and if any part of me gets into their brains, I promise to make them smart and fearless. Meanwhile, some part of me will come out of the other end to eventually find its way to the ocean, where it will be eaten by bacteria and plants. These will eventually be eaten by small animals, who will in turn be eaten by shrimp and fish, and I will be reincarnated into many different species. But this is not the end of my reincarnation. Some of the fish will be eaten by other fish, and some of these fish will find themselves on the dinner plates of many people. So I will become human again: many different humans from all over the world, male and female.

I will now be fighting among my many human reincarnations over food, territory and beliefs, and will also make love to myself through my many different reincarnations. I will go with my future human reincarnation to the Moon and to some of the planets in our Solar System, and may be lucky enough to travel into deep Space and to other solar systems. I will survive forever. I will be immortal because every time I am consumed by something, I will become it. The part of me that stays on land will fertilize the land and will be picked up by grass and trees, and will reincarnate into all kinds of greens, flowers and fruits. These plants will then be eaten by all kinds of animals, small and large, including birds. So I will finally lose my fear of heights, and will be able to fly without the need for an airplane.

I also will be reincarnated into all kinds of animals, some with excellent vision, so I won't need glasses or binoculars anymore. I will have better hearing than I ever did before, when I was alive in my current form. I will be able to smell the scents of the other far away male and female animals into which I have been reincarnated. I will also be reincarnated into animals that are active only at night, so I will lose my fear of the darkness, and best of all, I will finally have an interesting nightlife.

Oh boy! I cannot wait. Life really starts after death. The only sad thing is that none of my billions of reincarnations will remember me, because I surely do not remember where I came from.

So, if you want to be reincarnated, do not bury yourself. Recycle! Or at least donate your body parts. It is more fun.

37. Conversation with God on the Light Side

"Don't be possessive. Joke about everything.
You will be happier and will live longer."

-God-

In my sleep God whispered in my ear that He is about to close Heaven's gates because too many schmucks with no brain are knocking on Heaven's door, begging to get in. Heaven is too crowded; the rivers of wine are getting dry and God is running out of dopamine. People in Heaven party non-stop like there is no tomorrow (which is true; there is no tomorrow in Heaven). Some people in Heaven are complaining that their lives are unbearable. They do not have any challenges because all their wishes come true except the wish to go back to Earth where the food and dope tasted better and to see their loved ones. The only way to leave Heaven is to go to Hell and so they wish to go to Hell, where none of their wishes will come true.

The suicide rate in Heaven is very high. People are overweight and bored. They know everything and they are tired of seeing the same people every day, especially their mother-in-laws who invite themselves to dinner every day wearing only underwear. People in Heaven complain about not having privacy and they miss the real steaks and the real sex they had on Earth. There are no slaughter houses in Heaven. The cows and the chickens have their own Heaven next to the dog and cat Heaven, and they all live in peace and happiness. Humans and animals don't eat one another or any other life form. They eat pure recycled dung and imagination (Halal and Kosher), which are plentiful in Heaven.

God told me a story about this person who used to pray to Him at least thirteen times a day, begging for some kind of sign from Him. God gave Him no sign, and in his desperation for God's help, the man barricaded himself in his house and set it on fire. His neighbors tried to help him to get out of the house and he told them, "No thanks, God Himself will help." A policeman came and ordered him to get out, and the man said, "Only God can help me." And then the fire department tried to get him out of the burning house, and the man said, "Only God will save me."

God told me that after the man had burned and died, he came knocking on Heaven's door and complained that God didn't save him. God said, "I told him, 'I tried to save you on three different occasions and you would not listen.' So I sent him to Hell, where he is still praying for me. I think the man is just plain stupid. I should have made a monkey out of him! Speaking of apes, humans didn't come out of apes; apes came out of humans. All the humans who don't use their brains to their full capacity, or never get headaches, I make monkeys out of them after they die. So, respect the monkeys because they all used to be humans in their previous lives."

Why are some people stupid most of the time, and the rest are stupid all of the time? Because stupidity is infinite. It is always there, waiting for you. Why do the Jews use a lot of imagination? Because it is free. Why do the Muslims use a lot of violence? Because they think they are always right; their subconscious overrides their conscious mind. How do the Christians spread their gospel? They use lots of sticky honey.

God has a message to all the suicidal maniacs who want to take as many people as they can with them to Heaven/Hell while mentioning His name in the process--build the ultimate bomb-- God owns the doomsday machine--the black hole. You have to build a solid ball made out of a strong heavy metal, seven miles in diameter. Inside the ball you leave a small, perfectly round cavity that can very tightly hold twelve hydrogen bombs of one hundred and twenty megatons each. They have to be in a perfect circle and an exact distance from each other, around a ball of solid lead seven yards in diameter. The walls of the cavity in the large ball should be made out of stainless steel, with a mirror polish to reflect back as many photons as possible toward the center, where the lead ball is located.

Now the fun starts. You have to say, "God here I come," and pull the switch. If all the twelve hydrogen bombs go off simultaneously, and God gives you His blessing, you will create a black hole. The tremendous heat inside the ball will evaporate the ball of lead turning it into gas and plasma, and the tremendous pressure inside will squeeze all the lead atoms together to create a small black hole the size of a pinhead with a gravity force equal to the ball of lead seven yards in diameter. This is a small black hole with a very small gravity force, but nevertheless, it is still a black hole that will suck in any matter that touches it. The more matter it swallows, the black hole will become bigger and will have a stronger gravity force.

Within a few seconds it will swallow Earth, including its seven billion live people and about the same number of already dead people in their graves, including the fish (like Bin Laden), to become a black hole the size of a ball seven yards in diameter. Within a few seconds, all the people will line up in front of God for their judgment day. Most of the people will go to Heaven and some, including all the suicidal maniacs, will go to Hell for the rest of their eternal lives, without parole and without any clothes (it is burning cold in Hell). And they will all be wearing identical pig heads. This way, they will recognize each other only by the unique shape of their genitals.

In my honor God wants you to rename black holes "Eli's Holes" (for an out of this world description of black holes).

God told me there are no live people in Hell because the Ten Commandments in Hell start with "**THOU SHALT KILL**," and so the last nine commandments are irrelevant because there are no live people (only zombies) left to read them.

I asked God, "Dear God, God of all gods and Creator of our universe, when will you come visit us?" He said, "As soon as I finish building my flying saucer, I would like to come in style on a white donkey. Don't shoot me down." I asked, "God, dear God, God of all gods and Creator of our universe, when I am dead, will I know that I am dead?" "No stupid." (He called me stupid. I guess I shouldn't ask dumb questions.) "You wouldn't be there to know anything," he said. "Where would I be?" I asked. He replied that I would be everywhere and nowhere at the same time. (Now he was talking like a well-seasoned politician, saying everything and nothing at the same time.)

I asked God again, "Dear God, God of all gods and Creator of our universe, is there or was there **intelligent design**?" "No, stupid." (He called me stupid again. I guess I am really stupid.) "One more stupid question, and I won't answer you anymore. There is no intelligent design, nor was there ever one. But there is a **stupid design** and you belong to it. The proof is very simple. Nobody is perfect except God. You are always looking for perfection, without the ability to reach it." (God's answers are short, simple and elegant.) "You should teach your kids about this stupid design in schools, and also build monuments that last forever to glorify this stupid design. Join the crowd while you are still alive."

"Dear God, God of all gods and Creator of Heaven and Earth, I have a few questions for you about Heaven. Are there virgins waiting for us in Heaven?" "Yes my son." (He called me son. I guess He loves me or He feels sorry for me.) "The people who believe in me and worship me through the Islamic faith are entitled up to seventy-two virgins, the Jews up to sixty-nine, the Christians up to sixty-six, and the non-believers, well fuck'em, they don't get any (this "F" word usually has a negative meaning, but when God uses it it means unconditional Love or Love without boundaries). But are you sure you want virgins? These virgins are so ugly that they must wear veils at all times to cover their hairy faces and bodies. They have vaginas made from diamonds (diamonds in the bush) and your penis will be made from feathers. This way these virgins stay virgins for eternity. Are you sure you want virgins in Heaven?"

"How about women?" I asked, "What do they get once they arrive in Heaven?"

He replied, "All women, regardless of religion, are entitled to get up to seven Zohans, one for each day of the week."

"Can I see Heaven while I'm still alive?"

"Yes, by hitting your head with a brick or by taking mind-altering drugs like LSD."

In my dreams God gave me the friendly advice to tell my fellow Earthlings not to advertise Earth's location to the rest of the universe, its human population, and how advanced it is in science and technology. God told me that there are many advanced civilizations in our universe, but that they are paranoid. They will assume that Earth is about to occupy the whole universe, and will stage counterattacks against Earth. Besides, all those civilizations are flesh-eating, and human flesh is the most sought after meat, second only to pork, and planet Earth has plenty of both. In my dreams, I asked the God of Knowledge what the difference is between religion and science. God said that when your ass itches and you pray for relief through higher power, that is religion, and if you use your fingers for relief, that is science.

God promised me that I will be punished by death sometime in the future. My death will be caused by something that does not have a brain, such as a bullet, bacteria or a drunk driver. The reason for my punishment by death is my not keeping the Sabbath to its full glory. Sabbath, the seventh day of the week, should be a day of relaxation, a break with no work. But I let my heart work as usual on the Sabbath. I didn't give it a break. I never stopped it from working, and it was pumping hard every day. Every day was the same. It never got a break on the seventh day, and therefore, I should be punished by death.

But after my death, God promised me the most important job in Heaven: to be in charge of God's own sperm bank. I will have to make sure there is enough of God's sperm in the bank so that every mother on Earth will have a child in God's own image. This job is usually occupied by females (currently by Marilyn Monroe), who cannot resist God's glory and try to get themselves pregnant. No pregnancy is allowed in Heaven. Imagine your great-great-grandmother getting pregnant in Heaven. You wouldn't know your own family tree.

So, I will be the first male to occupy this job, maybe for eternity if I behave, if I resist the temptation of selling God's sperm on the black market. Because, after all, the sperm of the God of Knowledge is the most expensive, most valued, most elusive substance in our entire universe, just like God Himself. Praise the Lord, and pass the ammunition.

God told me that He is looking for a few good holy men to represent him and to spread his gospel on Earth, and in doing so, save Earth from its burning future (about four billion years from now). These holy men must meet certain criteria. These holy men have to be in good mental health. They cannot be holy if they were created through a sexual act. They cannot be holy if they had their mothers' vaginas around their necks. They cannot be holy if they kill and consume other life forms, including electrons, vegetables and chickens, for their own survival.

If you fit this criteria, please call God's secretary: 1-GOD-IS-GREAT (1-463-474-7328). If nobody answers, that means you don't fit the criteria. You are only human!

One last piece of advice from God is that if you don't like or understand this book, you may burn it. But if you do that, you will have a small brain and a large green nose once you arrive in Heaven!!!

www.ingramcontent.com/pod-product-compliance
Lightning Source LLC
Chambersburg PA
CBHW051437170526
45166CB00001B/22